Linda,
Th...
& hard work.
Spread Hope.

Love
Ann
2/19/19

Hope

THE DNA of HOPE

ANN-LOUISE JOHNSON, IFMCP, RN

Copyright © 2018 Ann-Louise Johnson

All rights reserved. No part of this publication may be reproduced, distributed, or transmitted in any form or by any means, including photocopying, recording, or other electronic or mechanical methods, without the prior written permission of the publisher, except in the case of brief quotations embodied in critical reviews and certain other noncommercial uses permitted by copyright law. For permission requests, write to the publisher, addressed *Attention: Permissions Coordinator,* at the address below.

Ann L. Johnson
1135 Georgetown Road
Suite 120
Christiana, PA 17509
717-786-4798

Ordering Information:
Special discounts are available on quantity purchases by corporations, associations, and others. For details, contact the publisher at the address above.

www.dnaofhope.com | www.annljohnson.com | www.wildheartstrong.com

Orders by US trade bookstores and wholesalers.
Printed in the United States of America
First Printing, 2017

ISBN - 978-1-5323-6439-6

NOTE TO READER:

This book was not written as a medical argument or to perch scientific jargon. Rather, it is a story about things that really matter—your life, your family, and your happiness. I spent forty years standing in front of patients in despair. And I found the medicine of hope to be as strong as any pharmaceutical development. With hope, many become victorious. I hope that after reading this book, you will join the ranks of victors.

THE DNA of HOPE

ANN-LOUISE JOHNSON, IFMCP, RN

THE HOPE CREED

Hope is bigger than the tallest mountain,

Yet small enough to live in me.

Able to calm the winds of sickness, and still the raging sea.

Hope is a pillar of sunshine, and no respecter of man.

Hope is fair, speaking childlike terms everyone can understand.

Hope reaches eight times into the soul.

Hope warms blood once cold.

And though others surrender to disease misunderstood,

Hope never gives ground, seeking the best over the good.

Hope will never touch a path previously trod,

But creates for the next traveler a manageable sod.

I believe in Hope.

Hope disrupts, and finishes the race.

Hope is committed, and maintains feverish pace.

So if in a pit you ever lay, your brow bloodied from the fight of the day,

Hope will carry your load when you've done all you can do.

And at the moment you want to give up on Hope,

Hope will never give up on you.

TABLE OF CONTENTS

INTRODUCTION I

CHAPTER ONE
Hope Never Gives Up 1

CHAPTER TWO
Human Touch 19

CHAPTER THREE
He Began to Lose Hope 35

CHAPTER FOUR
Rita 57

CHAPTER FIVE
Food Force 65

CHAPTER SIX
Sleep 95

CHAPTER SEVEN
Motion / Emotion 119

CHAPTER EIGHT
Support 133

CHAPTER NINE
Breath 157

CHAPTER TEN
Resilience 173

CHAPTER ELEVEN
Highways of the Mind and the Cell 195

CHAPTER TWELVE
The DNA of Hope Process 209

DEDICATIONS 233
ACKNOWLEDGMENTS 235
INDEX 237

INTRODUCTION

The Los Angeles Dodgers executive waited patiently in the clubhouse. He took a breath as their famed pitcher entered the room. It would have been acceptable to make a phone call, but he wanted to deliver the news personally.

"Tommy, your services are no longer needed," he told the respected player.

The veteran pitcher bowed his head, resting it in his hands. An awkward silence took over the usually noisy clubhouse.

Baseball was his dream. Baseball was his life. He'd been one of the most consistent pitchers in the majors. His work ethic was unequaled. His attitude is always positive, and his professionalism was a model for all to follow. However, an injury caught up with him. The Dodgers were signing younger, healthier talent. Tommy simply didn't fit the bill anymore. His time was up.

Nevertheless, Tommy learned a few things in his career. Struggle has so many benefits. Struggle teaches us to respect success. It shows us the value in hard work and uncovers the kind of wisdom embedded in a disciplined life. Through struggle, Tommy learned to never, ever quit. Having been in many battles on the pitcher's mound against formidable opponents, he learned that even in the bottom of the ninth with two strikes against you, there is still a chance. Most of all, he learned that in the midst of anxiety, disappointment, and fear, ask three questions. These three questions could calm any storm and provide hope when no hope was around. On that day, he took his head out of his hands, looked at his boss, and asked,

"Is there a chance?"

"Do I have a shot?"

"Is there something I can do?"

The executive, in an effort to be honest, smiled courteously and wished him well. Tommy had recently blown out his arm. His kind of injury was career ending. Sadly, he could no longer execute the bending motion needed to wind up and throw a major league ball. By all reasonable estimates, his career was over. No one had ever recovered from a tear to the ulnar collateral ligament. But Tommy, whose question was ignored by the executive, instead closed his eyes and asked himself,

"Is there a chance?"

"Do I have a shot?"

"Is there something I can do?"

Tommy then left the clubhouse and went directly to meet with an experienced surgeon. He asked the surgeon his chances of playing again if he had an experimental surgery on his pitching arm. This was 1974, long before many of the leaps in orthopedic surgery existed. The surgeon leveled with Tommy and told him his chances are one percent that he would ever pitch again. To which Tommy asked, "What are my chances if I don't have the surgery?"

"Zero," the surgeon answered.

"I'll have the surgery then," Tommy responded. "I'll give myself a chance."

Tommy asked the Dodgers for one favor: when spring training begins, please allow him to try out. In his words, "I have something the young players don't have; I have experience." Tommy continued, "If the younger pitchers are better than me, I will respectfully leave the field and never bother you again—but know one thing, for the next year I will be practicing every single day. When I come back, I'll be ready."

Tommy returned to spring training, tried out, and was re-signed by the team. He went on to win 164 more games with the Dodgers all because he refused to let go of hope. Fast-forward fourteen years later. Tommy was forty-five years old—the oldest of all players in major league baseball at the time. He was playing for the New York Yankees. The Yankees decided to cut Tommy from its roster to make room for younger talent. Tommy asked his team these questions:

"Is there a chance?"

"Do I still have a shot?"

"Is there anything I can do?"

Tommy won his job back and played four more years with the New York Yankees until he retired in 1989 as one of the longest playing pitchers in baseball history. He always had hope. He always gave himself a shot.

Today, the name of Tommy John is famous not just in sports, but also in medicine. In medical parlance, a Tommy John surgical procedure indicates you've torn your ulnar collateral ligament. Your doctor will take a tendon from another place in your body and use it to replace the torn ligament in the medial elbow in order to fix your arm. The procedure was named after someone unwilling to give in, someone who refuses to accept what life dished his way. He decided to live on purpose. He decided to perpetually hope.

It can be difficult to hold onto hope for your health when health professionals tell you your time is up. It can be emotionally debilitating when diabetes, cancer, emphysema, or any number of diseases begin to ravish your body. Where do you go? What do you do when your hopes of traveling the world or spending quality time with you grandchildren seem to be fading away? Ask yourself these questions:

"Is there a chance?"

"Do I still have a shot?"

"Is there anything I can do?"

And the answer to all of these questions is a resounding "Yes!" The days of medicine as we know it are over. Today, patients are now bold enough to participate in their healing. Brave enough to do more than listen to diagnosis, but instead, facilitate a new diagnosis through their actions. Having great healthcare is awesome, and having great physicians is always advised. However, neither money nor professionals can do for you what your mind and your belief can do for your body.

We all watched it happen. Billionaire Steve Jobs had everything, and what he didn't have, he created. Yet, out of nowhere at the age of fifty-six, sickness ended his life. Jobs had access to the finest, most innovative medical treatments known to man. There were no resources unavailable to him, yet he didn't survive. The human body often suffers when one assumes the only solution for its alignments must be found among synthetic creations—solutions by the hand of man. Mother Nature does know best. She knows how to fix what is broken, and she starts her healing at the cellular level.

There are trillions of cells in your body. Cells are the building blocks of life. Cells are the foundation without which nothing can exist. But when you think of a foundation, you think of permanence. Did you know your cellular structure has not only permanence, but also flexibility? Did you know your cell structure can transform instantly? Did you know your cells can regenerate outside of its kind? To activate your cells' amazing capacity, you must engage your eight genomic triggers of *support, motivation, food, motion, emotion, breath, sleep,* and *resilience.* When this is done, the human cell can bring into existence that which doesn't, in whole or in part, exist within itself. So why should you care? Does this have any real-world application outside of coffee table trivia? Yes, it does. If you have cancer, Alzheimer's, diabetes, or any degenerative disease, you must firmly understand your body is not only able to fight the disease, but can also create a new body within you. A healthy you inside of the sick you.

Your body can circumvent the current path of your sickness and build highways to health. This gives hope. But not only that, you need to understand that this very hope is in and of itself a cell. Hope isn't just belief or random optimism. Hope is the body's medicine. Penicillin for the soul. Hope is its own physiology—a tree of life with deeper roots than death. With hope, you can do the impossible. You can achieve greater sustainable health than you've ever witnessed.

Hope is its own example. There's no first mover needed. The magic of hope is this: one doesn't have to be exposed to hope to experience hope. Your body, your mind, and your spirit have ingrained in its very being all you will ever need to expect the best, most exciting adventurous living you can imagine.

I've found that most people are more comfortable going where they've never been if they have a map. Maps are evidence that someone has gone before us and memorialized the journey we now travel. So with map in hand, we replace anxiety with courage because we trust the intent of the map designer was for us to arrive safely. We drive highways at night because we assume the integrity of satellites haven't been compromised and so we are willing to, even with a newborn in the backseat, traverse mountainous terrain because we trust the system. A courage-generating system called GPS. Note how GPS doesn't require we have any personal knowledge of anything it presents to us as truth. We simply follow directions.

Medicine is very similar to GPS. Doctors, nurses, and practitioners don't have time to explain everything they know. They cannot transmit into our understanding all of the nights studying thousands of pages and highly intellectual tests required to put on the white coat. They engage us while we travel this journey called life. They assess where we are and tell us, from their elevated vantage point, how to get to our destination. Patients trust their knowledge and intent. Patients place their future in the hands of medical professionals. But there is one doctor you must trust *first* with your health before you ever see your primary physician. This doctor is your body.

Your body is the single greatest doctor on earth. Understanding this will reshape how you treat it, what you feed it, and who you allow around it. Once you go to see a physician, you are once removed from the most effective "healer" of the body—the body itself. We all know wonderful doctors and must always seek their highly skilled advice. But your body understands itself better than a physician ever could. Listen to it.

As you read this book, be empowered; be challenged; be bold. Allow these groundbreaking concepts of your eight genomic triggers to fuel your curiosity. Then challenge yourself to change your life, bending it toward health and away from disease. Finally, be bold enough to spread the news about the potential for all your friends and loved ones to do the same. Understanding your eight genomic triggers can end so many of your health problems before they have a chance to start. Most importantly, act now because this thing called life is far more unpredictable than it is short.

CHAPTER ONE
Hope Never Gives Up

On October 22, 2010, in San Jose, Chile, the first of thirty-three miners trapped underground for sixty-eight days emerged to greet family, friends, and prayerful bystanders. How did thirty-three men, with little food, water, or light survive being buried alive 2,000 feet underground without succumbing to despair? Hope.

They survived because they knew somebody was on the surface trying to save them. They survived because in their hearts was the tangible optimism of the human spirit. The understanding that no amount of dollars was too costly; no day was too long to seduce rescuers into giving up. Isn't that the message we want medicine to relay to all patients? That somebody is tirelessly working to save you, but there are logistics to saving people. You have to locate the lost person accurately. You have to assess the obstacles to reach them properly. Then you must design a plan —a tailor-made plan to bring them home to their family. Your health is no different.

Sickness can be like that dark mine those men were in. But knowing you aren't down there alone can be the difference between life and death. Hope is magical because in the absence of food, it can feed you. It is as tangible as the nose on your face, yet as unpredictable as the direction of the next breeze. And this is hope's true power—its ability to avoid inflexible definition, so hope can be to you whatever you need hope to be.

If you just still yourself for a moment, you can feel it. I can feel it. You can sense it. It's out here. It's everywhere. Turn down the noise, and you can almost touch the tangible vista right beyond your fingertips. Even if we can't express the feeling to those around us, in our quiet times of reflection, we acknowledge the inner pull, an optimistic reach for more—a desire to go further, climb higher, and enjoy life's adventures as a constant experience instead of random occurrences that rescue us from our day-to-day mundane existence. Somewhere deep inside, we know peace can be our live-in partner and not just a rare visitor. And it is hope itself that refuses to give you up. Hope refuses to let go, refuses to turn you over to the enemies of your health and happiness. You are the only one who can do that. Hope is stubborn. Hope is loyal. That whisper you hear when everyone around you says you'll never recover, that seemingly out of place sense of enchantment in the midst of chaos and discouragement —that is hope. Hope dares to speak up when all voices say there is nothing more to say. Hope is the last glimmer of light that leaves the eye. And in its daring audacity, hope changes not only how we feel about insurmountable odds, but literally populates our biological structure with courage beyond our own. When you can no longer fight, hope is your body's backup generator that will fight for you till the death. Hope believes in you.

When you find yourself in a free fall of life, if your health has failed, or maybe you're too exhausted to throw another punch, before you throw in the towel ask yourself:

"Is there a chance?"

"Do I still have a shot?"

"Is there anything I can do?"

And to all these questions, hope emphatically answers, "Yes."

ෆ

Health is such a part of whether we have quality of life or live in pain and discomfort. Every industry has taken leaps forward in making our lives easier, more efficient. The magic of technology allows us to communicate with people we barely knew existed just a few years ago, but for all our accomplishments, too many people still suffer unnecessarily from the same sicknesses and diseases from fifty years ago. Even with medicine's laudable achievements, more must be done.

It is with this understanding that we share with you The DNA of Hope. The understanding that the human body has infinite resources to destroy every obstacle of illness, disease, depression, sleep deprivation, obesity, and reduced performance. The knowledge that hope is our greatest ally, our strongest weapon of warfare against all we want out of our life.

At its fundamental core, The DNA of Hope is a healing web, a network of science, beliefs, and corresponding actions we explain as structure, function, and behavior. This book unravels the little-known hope network within the context of understanding how hope's cellular influence can aggressively change your future, just as your DNA's genetic footprint dependably reveals your past. The basecamp, the foundation of The DNA of Hope, begins with our cellular makeup. The DNA of Hope establishes its powerful position on one scientific fact: each cell in your body is a miniature you. Each cell in your body has in it the fingerprint, the blueprint to produce another you. We know this as parents, but generally relegate our ability to reproduce outside of ourselves when in reality we can reproduce *within* ourselves. You can reproduce a new you inside of you.

Once you digest you have creative influence over your cells just as you have influence over your children, you'll never look at your health the same. More health, more strength, and more time right here, right now with the ones you love is here within your reach. This book unravels this little known "hope network" within the context of understanding how hope's cellular influence can change your future just as DNA's genetic footprint reveals your past. To adequately appreciate the cell's ability to carry the creative power of hope through our system, let's briefly compare our body's processing power with the most powerful networks on earth.

The Most Powerful Processor in the World

Researchers in the Netherlands and the US call it "mind-bogglingly quick." They are referring to smashing the world record for data transmission over a fiber network. These researchers were able to push 255 terabytes per second on a single strand of glass fiber.[1] To date, the fastest single-fiber network connection can operate at a speed of 100 Gpps or roughly 2,550 times slower than what the researchers accomplished—2,550 times slower. The 255 terabytes per second achieved here is estimated to be more than the total sum of traffic flowing across the Internet at peak time. This is more data flow than millions of videos, billions of Internet searches, voice, video conferences, and large file uploads from every user across the globe happening at the same time. How were researchers able to do this? They did it by using something best understood as multi-core fiber or multi-core networking. The Internet today

---

[1] https://www.extremetech.com/extreme/192929-255tbps-worlds-fastest-network-could-carry-all-the-internet-traffic-single-fiber

operates on a single-core protocol, which essentially means fibers carry light from one single laser at a time. However, you can use wavelength division multiplexing (WDM) to squeeze more light down a single fiber. But that isn't a sustainable solution, so they moved to multi-core fibering.

Researchers took a glass fiber designed with seven individual cores and arranged them in a hexagon format. From this, they were able to hit 5.1 terabytes per carrier and squeeze 50 carriers down seven cores for roughly a total of 255 Tbps. By expanding the core from single fiber to multi-fiber, they created a data flow network that not only reduced the stress placed on one single line, but exponentially increased the capacity for the total amount of data transmitted. They increased bandwidth. This technological breakthrough is so new and so efficient that to implement globally would take billions of dollars to create systems, routers, and receivers to handle the data, not including years of workforce to implement with machines not yet created to install such a system. How does this innovation compare to the cell networking power in your body right now?

The biological complexity of the human cell network is nearly beyond comprehension. In 2013, The Okinawa Institute of Technology in Japan managed to create a computer network that simulated one second of human brain activity. This computational feat required the use of the fourth fastest computer in the world located at the Riken Research Institute in Kobe, Japan. The end result was creating an artificial neural network of 1.73 billion nerve cells connected by 10.4 trillion synapses. Admittedly, this is impressive, that is until you remember that medicine estimates the body has 80-100 billion nerve cells, which means the Riken Research computers only managed to simulate roughly one one-hundredth of our body's processing capacity, and this for only one second. For the fourth most powerful computer on earth to simulate only one second of brain networking capacity required 1PB (Petabyte) of system memory connected to 82,944 processors. Digest that for a moment. For context, the first sets of researchers were able to transmit 255 terabytes per second, which is faster than all the data currently on the Internet at peak time. One petabyte is equal to 800 terabytes. This means it would take approximately 261,120 terabytes to equal merely one second of the human body's networking speed. In the second it took for you to read the first five words of this sentence, your biological network processed more than every data transmission on earth. This is the power we have at our disposal. What do you think happens when hope is distributed among every cell in the most powerful network on earth?

These data numbers are overwhelming on purpose. A person dealing with serious illness is overwhelmed by the task of staying healthy. Overwhelmed by the possibility of leaving the ones they love, overwhelmed that those who have experienced the same illness have usually fallen victim to it. In other words, they lose hope. But when you realize what is on your side. When you realize the biological army available to fight for you. When you understand that there are more with you than with the disease, then an entirely different confidence is activated. Our bodies are self-healing, and the power of this healing is found in our cells—the biological army of our health. Our bodies have the processing power to destroy anything in it that isn't good for it. Our bodies have the processing power to turn positive thinking, positive eating, and positive perspectives into longer, healthier lives. Our cells have the power to take hope to every dark corner of our health and shine the light of recovery.

Understanding the newly birthed science behind The DNA of Hope requires you engage its opposition. When we say, "newly," this does not refer to science untested or recently discovered, but rather we refer to science that has been here as long as nature itself that only now can no longer be ignored.

Traditional Medical vs. Nature

There remains a general bias toward any practice not considered classically *scientific*—especially if this practice proposes to positively influence our health, Nutritionists and herbalists have known this for years. Even doctors of osteopathic medicine (DOs), which is the academic equivalent to the more widely known medical doctor (MD) classification, can attest to the ever so slight undertone of bias within the medical community against their holistic approach to patient treatment. Thankfully, irrefutable research is quickly eliminating such bias. Many can no longer ignore the power of proper thinking coupled with proper actions. Another point is appropriate here. Science is never created. Science is discovered. The brutal rigors of medical testing allow for dedicated scientists to discover what nature already knows. All that exists in medicine has existed prior to man's intellectual probing. For all our intellectual rigors, nature has been, is now, and will always be ahead of us. Partnering with nature is partnering with longevity—if we follow its lead.

This highlights the importance of proper perspective as we delve into sharing unique pathways to a healthier, more sustainable living regimen. This regimen combines the physiological and the psychological into one. No longer do we have to wonder why one patient full of optimism and hope, fairs so much better than another patient void of the same mindset. No longer is hope relegated to religious beliefs, positive incantations, and delusion. Hope is medicinal, tangible, and efficacious. Long before the practice of medicine existed, nature gifted hope with the capacity to instruct the human body at the cellular level, which is the basecamp where all the body's healing tools are stored.

Think of this practical example. Have you considered why your phone is called a "cell phone?" This is why. Cells are living organisms, and when telecommunications companies desired to communicate the power of transporting your voice, your instructions, and your life from one place to another, they aptly titled it "cellular service." Interesting, isn't it? Cellular companies borrowed a medical term, a biological term to articulate their commercial service. Why? It is because they understood the functions are the same. Your cell phone is an extension of you since it allows you, your voice, and your desire to travel with you wherever you go. Your cell phone does what you tell it to do. It allows you to speak to whomever you wish and accomplish much more than you could if you were bound to one location. Smart phones do even more. The more apps you put on a smart phone, the more information it returns to you, and the easier it makes your life from keeping up with your finances to your fitness. So then, your cell phone is basically a blank palette onto which you can paint what you need it to do for you. It goes further.

When a person is lost, even when incapacitated and unable to call for help, their cell signal can save their life. In other words, what they have done prior to becoming lost is what makes sure they are found. Turning on the phone puts the cellular towers in play as they are constantly sending and receiving messages. And this is *precisely* what hope does.

Hope is like a biological cell phone communicating to the biological cellular tower (the body). Hope not only identifies where you are, but where you need to be. Hope gathers the health forces at the cellular basecamp in your body and gives instructions these forces on how to save you. Hope saves. This represents a seed change in our collective thinking. Hope naturally reaches toward the future; while DNA naturally reaches into the past. When we network the two, we engage the power to create the life we want using eight genomic triggers: *support, motivation, food, motion, emotion, breath, sleep,* and *resilience.* Hope is the lynchpin between what we've already had and what we want.

To fully enjoy life, you need to be healthy, and being healthy requires information, courage, and most importantly, support. The first thing a paramedic does upon arriving at the side of a severely wounded patient is to give touch, to transmit hope from hand to hand—to transmit hope from voice to voice. "What's your name?" is a standard question asked of the accident victim. This is no random inquiry. Reminding us who we were in better times encourages us to grab hold of the potential of such better times continuing. It is hope building a bridge from our past to our future while hovering over our present circumstance. Our name reconnects us when we feel isolated by painful events. The touch communicates with our cells who know when they've been connected to healthier cells. The hand of the paramedic, the voice of the paramedic, has saved many lives. Hope works. And in the end, that is all patients really want to know—what works.

The Picture of Health

Think of your health picture as a jigsaw puzzle. Hundreds of pieces lay in front of you. One piece is labeled "Food" another reads "Sleep." Random pieces in the corner read "Motivation," "Support." And next to these are even more stacks, "Resilience," "Emotions," "Breath," and "Motion." Every imaginable part of your health lay scattered on the floor. The shapes are mismatched, and the illogical color scheme makes more complicated an already tall task of organizing a picture you've never seen before. You have no map, blueprint, or instructions as to how and where you should begin assembly, but your future depends on doing this right. And so, frustrated, you—like many before you—take the only sections you understand and piece together the rest the best way you know how. But what if you had access to clues? What if you were given eight genomic triggers covering how all of these factors interconnect? Triggers that made sense of all the disjointed pieces? I've just explained something called functional medicine. A fresh approach to your healthcare that reveals the missing pieces of information you need to take control of your life. After all, it is not a puzzle, it is your life.

Living healthy is not as simple or as complicated as it seems. Not as simple because there are so many expert voices telling you to go right, and as soon as you start going right, another chorus of experts says go left! What do you do? You do what works for you. And this hits at the "not so complicated" part of living healthy. When you walk into a clothing store, no matter how beautiful the clothes are, you only buy what fits you. The same applies to your health. It must be tailor-made, it must be functional. Your health plan must fit you. To find what fits you, a few critical points must be understood. First, once you remove genetic conditions forced upon a person at birth (down syndrome, sickle cell anemia, etc.) or the malnutrition suffered by millions in third world countries who have little chance of changing their fate, what you're left with is people dying from modifiable lifestyle factors. Factors such as diet, overweight, inactivity, toxic environments, and toxic people. In other words, most people die by choice, not by force.

Second, we have the research to modify these factors, so that they do for you what they usually do against you—that is give you life, not take it away. Inherent in each of these modifiable factors resides what we call The DNA of Hope. The DNA of Hope is best described as the reproductive system of wellness. We all remember in elementary school first learning of DNA and the barbershop-like strand representing its cellular body. Well, think of The DNA of Hope in a very similar fashion, except this time the double-helix strand are instructions for wellness, for healthy living, and stronger thinking. All to give you more life, more strength, and more time. And to ultimately reproduce itself in the lives of your offspring just like genetic code. Third, and most important, the courage required to change your life doesn't have to be your own. What do I mean by this? You can engage a courage-based system that will go ahead of you and accomplish for you what you may not be able to accomplish on your own. In this system, the need for "human touch" is eliminated. This means an approach with no dependency on human interaction for the end objective to be achieved. The goal is for any person, of any skill level to be able to engage principles of The DNA of Hope and accomplish the same results. The non-scientist is now able to access scientific information. The disadvantaged now has the advantage.

☙

Even the most complex of subjects are best relayed using simple stories. The magic of story can never be fully quantified. This is because every story told meets a story lived. Successes, failures, dreams unrealized, and silent regret creates a filter through which we view the world. This makes it critical to understand the word "patient" doesn't connote a number, but an individual, a life, a story. Sidney MacDonald Baker, MD, explains the critical importance of understanding the uniqueness of patients this way,

> *The logic of complementary, functional, integrative, or good medicine begins with the biological reality of individuality and proceeds with the following strategic models that can be posed as questions.*

Is there something for which this person has a special unmet need? Is there something to which this person is intolerant?

Baker goes on to highlight an unyielding redundancy in the fundamental areas of the unmet and intolerant places of individuality. Areas of lack or fullness have the same cycle. Internal good tends to reach outwardly finding relationships and habits to fuel the nature of its core to sustain its existence, areas of bad tend to find poor relationships and unhealthy habits to also keep itself alive. In the end, good and bad equally want to live. We, the medical professional, can understand the good, the bad, the unmet, and the met needs of the patient through listening to their story. The patient's story points the way—if we are willing to listen. Their story tells us which triggers are strong and which triggers are lacking. The patient story reveals whether they need more support, better eating habits, emotional stabilizing or maybe better breathing habits. To the level we listen is established the level we learn. And what we learn—we believe.

Belief is the last critical key needed to open the door for hope to come in and partner with our incomparable cellular structure. Belief is the nutrient of accomplishment. Belief is the enemy of "can't." This seems to be a rather logical statement, but logic isn't your friend when it comes to putting down a boring life and picking up one of adventure. Our logic, or the accepted order of thinking, grows out of what is shown to us, taught to us, exampled before us. Belief is powerful because it closes the gap between what we see with our eyes and what we see with our hearts. This presents a unique challenge. It can be easier to dream bigger dreams for others than we can for ourselves. Imagining success for "other guy" is more "logical" because we don't intimately know his flaws, fears, and areas of doubt. This is one of the reasons people who are sick can usually envision someone else getting well before they can envision this for themselves. Ask a sick person about their sickness, and they will relay intimate knowledge of who in their family didn't survive from sicknesses X, Y, or Z. In the end, negative expectations can prevent us from taking the very action that would save us and separate us from the suffering pack—even if the suffering pack is our own family. Now, positive expectation isn't a sole determinant of an event, at least not in ways that are consistently, scientifically provable. That said, day-to-day life is a direct result of what we expect it to be. Henry Ford was correct, "Whether you believe you can do a thing or not, either way, you're right." Do you believe you can be well? Negative events don't require belief. Isn't that interesting? I once read in a book that weeds don't need fertilizer. This makes it that much more important to feed yourself with affirmation, love, healthy relationships, and healthy living. Give yourself a shot.

Our experiences form our belief system—a system that's nearly impenetrable. This is because what is personal is true. No one can tell you what you felt when you felt it. No academic research on human touch can simulate human touch. We are experiential beings. This makes critical the expansion of our experiences so we know more than what's possible intellectually, but experience these possibilities through our senses. We have no other way to interact with the world but through touch, smell, sight, hearing, and taste. And I'll add an extra sense from our earlier conversation: thinking.

Thinking properly allows us to, in the absence of personal experience, take on the wisdom that would have been bestowed upon us had we personally experienced an event. Imaginative thinking can transport the lessons of another into our lives. Imaginative thinking allows us to gather fields of grain without ever walking the field. Imagine yourself healthy.

Nothing is more debilitating—nothing is more depressing than being sick. It is as if the world is on the other side of thick prison glass. You can see everything, you can faintly hear the laughter, but you can't participate. It is easy to forget the obvious; the point of health care is to help people. No one cares about public policy, biotechnology, or pharmaceutical research when you're holding the hand of a loved one—when their grip is loosening as their vital signs drop. Medicine has no more sacred mission than to push that painful day as far out into the future as humanly possible. Above all, hope.

CHAPTER TWO
Human Touch

Peter spent all day roaming the countryside learning of the people of Calcutta, India. Time flew. Being late to rehearsal was guaranteed, but his gut said stop, and so he did. Ten minutes later, Peter was back on his way to the rehearsal hall. He paused, turned around, and with violin in hand, gave a woman one last parting look.

"God bless you," he said, crossed the street and raised his hand for an approaching cab. At that very moment, from behind the dense arrangement of merchant tables, curtains, and melodic trinkets, a voice asked him one question that would change his life for a second time. He thought he heard the question incorrectly because it simply didn't make sense.

"I'm sorry?" Peter asked.

The question came again.

"When did you die? Where was it? Who was with you in your final days?"

Peter took one step closer to the frame of an elderly woman

speaking in his direction.

"Who touched your forehead with the back of their hand?

Who wiped paste from your lips in compassion? Who son?"

"Excuse me ma'am?"

"I must know the name of your friend," the woman replied.

Peter simply didn't know what to say. He looked down at the woman's shaking hands. They seemed to be wrinkled by time, but not chronology, the sort of time spent learning not just living. She never broke her stare, her one good eye attached to both of his.

"Son?"

"I'm right here in front of you ma'am," answered Peter.

"With all due respect, I'm not...well, I don't understand your question."

The woman smiled.

"Son, a moment ago you walked up to that very sick girl over there." The woman pointed across the dirt road smoked by the congestion of passing cars on a road meant for pedestrians.

"You turned around and placed money in her hand. You didn't drop the money, you grabbed her hand, son. Her back is full of sores but you knelt down and hugged her. Her lips are cracked with hunger, but you put your face on hers. Her hair is a nest for flies. Blood shapes her footprints. No one speaks to her, son, no one, not even the priests. Why did you? Why did you do that? Tell me when you died."

Peter's confusion deepened and heightened at the same time, but the woman wouldn't let up.

"She used to be a beautiful young girl. I watched her as a lad." The woman began to cry, her tears cutting a clean path through thick dust on her face. Her voice now barely a whisper, she asks Peter again.

"When, son, when was it? Please tell me."

Peter slid the violin strap off his shoulder, knelt down and nervously whispered in return.

"Why are you asking me this?" Peter said.

"I'm here in front of you. I don't know what you mean. I mean…I just don't know how to…

She interrupted.

"I'm asking when did all of the inhibitions, the fear, the arrogance of good health, and the thumbing of your nose at the stench wreaking from her body go away? When did you become alive son to the pain of others? When did you become a caring human being willing to risk your life to comfort another? A person has to die first before that happens—the kind of death that allows you really live."

Silence.

Peter's frame became a mannequin. Her words seemed to stop time. A deep warmth overtook him as if he were separating from his body, and then his thoughts rushed back to that hospital bed fifteen years earlier. You see Peter himself had been sick unto death. Every doctor, every specialist from every hospital gave him no chance to live, but miraculously he recovered. He hadn't made a conscious effort to do what the elderly woman observed from a distance. He just knew how it felt to be shunned when his body was in indescribable pain and his heart in despair. It was in this most terrible moment of his life he felt something. The squeeze of a hand. The first thing he noticed is the hand didn't have a sterile glove on it. It was warm; it was human. Then came the soft, yet authoritative voice from a total stranger.

"Sir, what do you love the most?"

Peter was so sick he couldn't turn his eyes enough to the left or right to see the person behind the voice, but the question caused him to reach for enough breath to answer.

"I, I love playing the violin." Peter said in broken voice more characterized by tremble than tone.

"I love music. I wanted to play in orch..." Peter squeezed his eyes tight, wincing, as another lightning bolt of pain shot through his body.

"I wanted to play in orchestra one day. That is—was—my dream."

"Yes!" the stranger said. "You are a great player. I can tell by the way you talk about it."

Immediately, Peter's chest jerked upward, his eyes confirming what his mind knew all the while. He would never see his dream come true. His time was up. And then the woman, whom Peter had never met before and for whom he had no name, placed her other ungloved hand on his shoulder, moved her mouth to his ear and whispered.

"You will play again Peter." Instantly, Peter snapped out of his memory to find the elderly woman on this Calcutta sidewalk staring at him with the same confusion he stared at her when she asked her first question. She recognized his thoughts had gone somewhere else, somewhere meaningful.

"I...I understand your question now." Peter said. "I know all too well how that young lady feels over there. I stopped because I wanted to give her what was given to me in a hospital bed fifteen years ago. A woman I'd never met gave me hope. I'm here today playing with an orchestra on a mission to bring the sounds of hope to impoverished nations. We are called Hope Philharmonic. My name is Peter."

"I know," the woman responded.

"How do you know my name?" Peter asked.

The woman smiled, raised her handkerchief to her eyes and squeezed a wink. Peter looked down at his watch. He had more questions but needed to go. He restrapped his violin case and hailed a taxi. That night, the Hope Philharmonic played for a large gathering of people—poor people, sick people, people with no hope needing to hear what joy sounds like. Peter was there for them because fifteen years earlier a strange woman was there for him. When someone shows us they care, it forces us to care for others. Hope is generational in that it always births offspring, just as in traditional DNA. You can trace a person's good will back to the good will of someone in their life years prior. This, among other reasons, is why hope matters so profoundly. It is regenerative in nature.

The story of Peter points to the unavoidable prerequisite for activating the hope network: motivation.

It was motivation that moved Peter to pick up his violin again. Raw belief from that stranger in his hospital room transferred to Peter the heart of "I can," just when he thought he could not. He became motivated again to accomplish his dream after he had resolved it was gone forever. It is interesting how many times we are pushed to do something great by someone with whom we have no personal relationship. Intimacy isn't necessarily the place where we discover our greatest breakthrough. Often, a visionary word of encouragement from the mouth of a stranger or from a book is powerful enough to connect with a place usually unreachable by those outside of our relationship comfort zone. There is no greater example of this than the healthcare professional.

There's no location riper for motivation than the sacred confines of a hospital room. We go there for a reason. We are attentive and absolutely open to the suggestions given by the professional in place to treat our ailment, but no matter how effective the information given to us, how we view this information makes all the difference in the world. The problem with problems is they usually seduce us into looking elsewhere for the solution, when, in reality, all problems have their solution built inside of them.

In 1945, psychologist Karl Duncker created a problem. It is famously called The Candle Problem. Duncker brought a group of individuals into a room and presented them with several items—a candle, a box of thumb tacks, and a pack of matches—all sitting on top of a table. He then gave one simple instruction: attach the lit candle to the wall so that it doesn't drip wax on the table. Many people attempted to thumb tack the candle to the wall, but that did not work. Others tried to melt the sides of the candle and then stick it to the wall, which also did not work. After ten minutes, the answer was found in overcoming a behavioral tendency called functional fixedness—a cognitive bias that prevents many from seeing creative or extra-logical solutions. Ultimately, the answer was to thumbtack the box holding the thumbtacks to the wall, placed the lit candle upright in the middle of the box so that the melting wax collected in the box and didn't touch the table. As soon as the participants looked at the box as more than a receptacle for tacks, the answer became obvious, but as long as they believed the box to have a single usage, the problem could not be solved. Motivation is no different. You must believe motivation can do more than what you currently believe is its primary purpose.

If you believe motivation is only useful to perform better at your job to receive a year-end bonus, or that it is simply a form of temporary inspiration that puts in motion what an uninspired person wouldn't normally do, then that will be the extent you will benefit from this most powerful genomic trigger. But if you realize motivation has not only psychological but even greater biological consequences, if you believe that motivation can not only change your reason for living but give you more life to live, you are beginning to understand its relationship to the cellular composition of our bodies. You are beginning to view it as the box in which to catch the dripping wax.

In this book, motivation is the second genomic trigger in the network we discussed earlier. Motivation activates our biology. Motivation starts us moving, living, and thriving, or starts us dimming and dying. If you look at motivation as a function of harmony versus dissonance it is easier to understand its profound effect on our cellular influence. Harmony, defined by that which is in place with another thing or person, is the natural order of things while dissonance, even when forced, refuses to bring peace to the ear, mind, or body. Motivation is inherently in harmony with our body's cell structure.

All of the body's functions are designed from the perspective of harmony. Chromosomes come from our parents in harmonious pairs—forty-six together, twenty-three from each parent. On these chromosomes live our genes, over 20,000 of them. These genes make proteins and other things. These proteins are made up of amino acids and amino acids do the work deep within the cell by communicating and sending signals like the cell phone and cell tower example we discussed. Amino acids amplify messages, decode messages, and regulate reactions by making enzymes. They build and repair cells and tissues. In short, they make up the behavior of the cell. It's a fascinating biological operation that exemplifies efficient instruction and obedience and all of this is triggered—positively or negatively—by motivation.

Nature is intrinsically "motivated" to work. That is to say, it is naturally motivated to maintain equilibrium without outside force. Man does not have the power to improve or systematically nudge nature as its forces are perfectly balanced. Wisdom is to discover nature's genius and live in harmony with it. When we do, we are rewarded with life. This is best exemplified by the basis for all life: water, without which no man can live. Comprised of two atoms of hydrogen and one atom of oxygen bonded together, water, without trying, never loses its chemical bond. Why does this matter? This matters because when you speak of the hope network, cellular structure, amino acids, proteins, enzymes, and how all of this works together, you must know what you are really saying. When you consider what has been proposed in relationship to motivation as a genomic trigger, you are making a claim. You are saying these things are cemented in nature, eternally unchangeable, but also eternally flexible. That is, cellular biological functions are as open to modification as much as water is closed to modification.

The Science

The old-world way thinking wrongly concluded that one protein was coded by only one gene. Today, on the horizon is a new understanding of a grand autonomy within the cell. We understand now that DNA within the nucleus is copied and made into a premessenger RNA. Before the premessenger RNA leaves the nucleus on its journey to make amino acids, an additional cut occurs. We call these cuts introns and exons.

When parts are cut out, they are called introns. What is left within the nucleus and the remaining parts stitched together are the exons. What is the purpose of the intron and the exon, and why should we care? Introns and exons are what add variability, flexibility, and plasticity to your cell.

Furthermore, recent studies reveal a specific gene called

CSLO. CSLO uses alternative splicing to create more than 500 different proteins, each with a slightly different electrical response.[2] When researchers studied the hair cells on one end of the cochlea, they found one of the proteins. When they looked at the hair cells on the other end, they found another protein. In between, they found 574 other proteins, each of which was ideal for producing a strong response to a specific sound frequency. In other words, the exons and the introns of the CSLO gene had been set up so that each hair cell can use an alternative splicing to produce a protein that is perfect for the frequencies to which it is supposed to respond the strongest. This is why lower frequencies of bass sound like bass to you and higher frequencies of treble sound like treble. Researchers at the Mount Desert Island Biological Laboratory in Salisbury, Cove, ME, and at Stanford University's Department of Medicine have identified a parallel human gene to CSLO called hSlo. The following is taken from an article published online in 2015 on the subject of subunits and splicing. It reads in part,

> ...cSlo, a version of the Slo gene sequence (an older term for the BK channel subunit) found in the chicken; hSlo, a version of the Slo gene found in humans; Kcnma, another term for the β subunit of the BK channel; Kcnmb4, β subunit number 4 of the

---

[2] https://www.ncbi.nlm.nih.gov/pmc/articles/PMC2897565/

BK channel; KCNN3, one of the sequences for small conductance calcium-activated potassium channels... [3]

This quote means that nature's design is so genius, so detailed, that each hair cell can choose from among 576 different proteins in order to deliver the best frequency response for your hearing pleasure, all as the result of a single gene. Now what we have, instead of one gene making one protein, we have located variability, plasticity, and flexibility within the human design. In essence, you have 25,000 x 576 different proteins at your disposal. Try multiplying 25,000 x 576 and you come up with "new think." You come up with a number that shatters conventional thinking of one gene, one protein.

With this, variability is human—flexibility is human—and this reaches all the way down to the cellular, biochemical, genetic level. We continue to learn about our genetics and our genomes. We stand amazed at the power we have inside of us, but are we using this power, this information, to *motivate* us toward a healthier, stronger, longer life? We haven't yet. Not until today.

Take an adventure into the cell, the cell wall. This is where motivation pushes our bodies to their untapped capacity. We have the power to turn motivation on or off. It is this deeper science that propels us right into longer life. Our ability to

---

3 http://www.sciencedirect.com/science/article/pii/S037811191501481X

recognize the open architecture of the cell is monumental, and with motivation as our propulsion, we can dive directly into the cell, into its flexible components, to create the life we want. The cell is our foundation on top of which all life gets its footing. The cell, then, provides a magical blank canvas allowing us to paint excitement, adventure, and our very best performance.

Motivation obtains much of its power from unrealized dreams and unreachable aspirations. Motivation can be fueled by the mundane or by critical realities of survival—like the need for food and water in a barren landscape far from civilization—but motivation is most beautiful when it takes on the form of words. Words, that would otherwise be empty, fill themselves with healing salve when spoken by one with support in their spirit. A simple word of encouragement can move those about to give up to hold on.

☙

The final touch of sun disappears from Calcutta, India. The various musicians from around the world that make up Hope Philharmonic pack their instruments, preparing to move to the next town. But one player, just one, has a small crowd gathering around him. Maybe it was the violin solo he played while interacting so passionately with the audience, or maybe it's just his magnetic smile; whatever it was, Peter was the star tonight.

His playing inspired all in attendance. A collection of boisterous children surrounded him on the side of the stage. Teenagers, adolescents, and young children were cheering while others sang nursery rhyme melodies. The youngest ones pulled at his pant leg as only innocent children can do. Sheer joy and thankfulness was in the air. Exposure is an adventure all unto itself, and tonight music took these children on the greatest adventure of their lives.

"Peter! Peter!" the children chanted. "Peter! Peter!"

To have musicians bring cheer to a cheerless countryside has lifted the spirits of an entire community, but none of those in attendance will ever truly understand the power behind Peter's playing that night. None of the children understand how close Peter was to never playing again. They will never know how sick he was in that hospital room fifteen years earlier. They will never know the story behind his smile. The sound of triumph that now rings from his instrument isn't the sound of years of practice, but rather the sound of overcoming despair, the sound of escaping unrelenting pain—the sound of a second chance.

That woman who whispered in his ear a decade and a half earlier planted in Peter's soul the fire to recover. Her belief stood in for his when he hadn't the strength to see himself healthy again. Where he saw an ending, she saw a beginning. She pulled him away from the edge. Peter's eyes threaten tears as he looks in the faces of all the precious young lives he must now leave behind. He knows the only difference between a life of fulfilling your dreams and one of pain and suffering comes down to being motivated to do better by someone who sees better in you.

The crowd thins and Peter is stepping onto the bus to head back to his hotel when a small, tender voice catches his ear.

"Sir, when I'm big, I want to play the violin."

Peter looks down. He sees a little girl barely clothed. A torn piece of linen is all that covers her bony shoulders. Trying to hold back his emotions, Peter whispers an answer in a reflective tone.

"You will play one day, young lady. You will play."

CHAPTER THREE
He Began to Lose Hope

Hope often emigrates itself into our lives through the passageway of unsolicited motivation. This is why the people, places, things, and ideas we entertain are so vital, as hope and hopelessness are often given to us without our permission. It is then motivation that creates capacity to hope when our personal desire to do so has been exhausted. Exhausted from failure, discouragement, or even previously held hopes that were unconsummated. Unconsummated hope is a consequence of timing rather than truth. For this, look no further than the founding fathers of numerous corporations worldwide. Many began their companies in hopes they would become the very companies they are today. Yet, their lives ended prior to realizing this hope. Today, heirs and strangers benefit from hope originated by another person in another time. This teaches us that hope, like energy, is never lost, only transferred.

Motivation has the power to give you power when you are

at your weakest. It is fuel for the eternal engine of hope. As we learned with Peter in the last chapter, those who motivate us cement their positive influence into our lives, and, by doing so, transfer to us a moral debt only satisfied by motivating others for good. Motivation is magical because of its extra-logical tendency to strengthen our grip on the tattered rope of unmet expectations. Motivation reveals the equality of fatigue that awakens us to one powerful fact; it will take as much effort to win as it does to lose—so choose the former. When you stand at the edge of a broken bridge unable to return to the comforts of life before, only then does motivation enable us to do at one time in one place something we may never be able to reproduce again in our lifetime. Just ask Harrison Okene.

It is May 29, 2013. Rescue scuba divers were given the order to recover bodies from a sunken vessel. What was at first a rescue mission became a mission of recovery. Uncertain what they would find, the divers took flashlights, ropes, and harnesses to pull whatever and whomever they found back to the surface. They also took a video recorder to investigate how and why this tragedy had occurred, but they could never have imagined what they would end up getting on tape.

Only three days earlier, a tugboat carrying twelve souls sank in the middle of the sea in the middle of the night. Understanding the area where the vessel was lost, officials notified families as to the certain fate of their loved ones—loved ones who fell victim to unforgiving waters at 5:00 am off the coast of Africa. The small, dedicated crew had been towing oil in these early hours when an ocean swell slammed into their vessel, snapping its tow rope in half. The force sent the men to the bottom of the ocean without warning. A man named Harrison Okene was on board. A great cook, Harrison took the trip to provide solid meals for the crew. His decision to go would prove to be fateful.

At the time of the accident, the entire crew was asleep, locked in their cabins. This was a safety precaution. In this area of the sea, pirates were known to board and harm crews in the middle of the night. Harrison would have been locked in his cabin also, but he needed to use the restroom, so he got up at around 4:55 am to do just that. He was still in the restroom when the rogue wave hit. He found himself being tossed up and down so violently he was nearly knocked unconscious. The crew members awakened injured and disoriented, but were unable to get out of their locked cabins before rushing waters drowned them. For no apparent reason, Harrison had left the door to the restroom slightly open. When the wave hit, the force catapulted him into the hallway. The hallway was quickly filling with water, and he began to swim frantically. With only seconds left before needing to inhale, his hand pushed on a door to the engineer's room. The door opened, and momentum pushed him upward against a ceiling.

Gasping, Harrison noticed something odd. He could breathe. Somehow, the engineer's room had an air pocket. What Harrison didn't know was that every room, every space, in the entire boat was filled to capacity with water except the one he was in. Mysteriously, the room had two feet of space left where no water had invaded. This was not only improbable but illogical because the room he was in was at the very bottom of the ship. Harrison took a deep breath, and with despair creeping in, recognized his fate. He didn't know how much time he had left, but this small window of opportunity would allow him to make peace with his maker. He replayed his time with his wife, his family, and his friends. He thought deeply of his mother. His thoughts were interrupted by terrible, sharp noises. Sharks had reached his crew mates. Harrison decided he wouldn't go down without a fight. He grabbed a metal object to give himself a chance if the sharks came his way. He waited and waited. In between sounds of terror, he considered what he could leave as a last sign for those who may find him at a later date. He closed his eyes and begged God for help. When he opened his eyes, he saw something floating in the darkness of his confined space. A plastic bottle of Coca-Cola.

It was just floating by as if nothing was wrong. He grabbed the unopened bottle. Somehow, in some weird way, the Coke gave him a tiny glimmer of hope, and when you're 100 feet underwater with no chance of getting out, any sign of how your life was before you were in such a desperate situation can motivate you to hold on just a little longer. Sometimes, all we need is just a little longer.

Three days later, the rescue scuba diver had located the ship and was able to communicate with officials on the surface that the vessel was indeed resting at the bottom of the sea. What Harrison suspected was correct; it seems sharks had torn into many of the rooms and devoured many of his crewmembers. Using his video connection, the scuba diver motioned for his team to pull him back up to the surface so he could report what he'd found. The diver's flashlight grazed over a small port window. And through the dark, polluted, dusty water, a human hand could be seen.

"My God!"

The scuba diver was frightened. The hand was withered and wrinkled but moving. The hand was reaching for the light and pressed its palm against the window. The scuba diver moved closer and raised the beam of light higher to see if he was just imagining the impossible. His light lit Harrison's face. He was barely alive. The scuba diver panned the camera further to see the other hand holding the bottle of Coca-Cola from which Harrison had been taking sips of soda and sips of encouragement. The otherwise useless 99-cent plastic bottle helped motivate him to remember hope can take any form at any time and be for us what we need it to be.

The diver radioed the amazing news back to the surface. There was one survivor alive at the bottom of the ocean three days after the sinking of the ship! At the time, no one could take the time to process the unbelievable nature of this news. They needed to get Harrison to safety. After his successful rescue, and after many hours of depressurization and medical checkups, he decided to speak of his ordeal. One particular phrase stood out from the others. He said, "I began to lose hope."

A gem is buried in his comment. His comment proves he had hope in the first place. The details of how he survived also imply that something as meaningless as a soda can reset the mind to the possibility of making it out alive. That is what hope does. When in desperate circumstances, hope scans our atmosphere, our real-time thoughts and past memories. Hope takes in what we hear and what we see to find the smallest seed of positive thought. And with this seed, hope expedites the normal process of growth and blossoms oak tree-like encouragement to shade us from the despair of our current circumstances. And this tree-like encouragement during desperate times takes on the same form as hope's internal movement along strands of the human DNA structure—spreading, reaching, inhaling and exhaling life, adventure, and belief. With deeper roots than death and higher heights than the heavens above, hope—and hope alone—can leverage its ability to rewrite life's firmest conclusions of physiological requirement for survival. Harrison said he started to lose hope, but thankfully that's as far as it got. He never lost it.

There's an interesting side note to this story. Human beings inhale roughly 350 cubic feet of air every twenty-four hours. In commenting on the miracle of Harrison's survival in an online physics forum, physicist and scuba diver Maxim Umansky at the Livermore National Laboratory estimates that his air pocket was effectively compressed by a factor of four. Carbon dioxide, which is deadly to humans at a concentration of about five percent, should have been a real danger to his survival, yet he survived.[4] Umansky comments, "A human needs ten cubic meters of air per day. That guy somehow survived for two and a half days." How does a man at the bottom of the ocean without any way of communicating, stuck in an upside-down ship with water up to his neck survive for three days holding on to and sipping from a bottle of Coca-Cola? Hope has the ability to modify science itself, causing the body to survive on less nourishment, less oxygen, and less heat to give you just a little longer. Sometimes, all we need is a little longer. But here's the best news of all: Hope isn't just to rescue us from desperate situations. In fact, the real power of hope is found in times of peace. Hope isn't a tool for an emergency; it is a tool for a lifestyle.

After caring for thousands of patients, I've learned one thing.

---

4 https://physics.stackexchange.com/questions/67970/surviving-under-water-in-air-bubble#comment137400_68008

Part of the entire process of finding better health is finding what ultimately motivates a patient at their highest level. When you find such motivation, the patient is able to access peace in life and in death. There is a sobering scene in a movie called *The Grey* starring Liam Neeson. Injured in a horrible crash, one of Neeson's colleagues is losing blood too quickly to be saved by a tourniquet. Neeson, seeing the fear on the man's face, kneels down next to him and touches his frightened skin. He asked his colleague to think of his daughter. In all their years working together, nothing brought happiness more than when his colleague spoke of his little girl. Now faced with an insurmountable reality, it was this love that motivated the man to leave this world in peace by reminding him of the only person able to overcome the knowledge of his pain and of his plight. Memories of his daughter were greater than signals sent by his nervous system to his brain. Signals of extreme pain, fear, and despair were overridden by a greater signal—love. If you step back and really investigate how something like this can happen, the medical implications are profound.

They are profound because pain is our body's ultimate alert system. Pain is loud. Pain is rude. Pain lets us know something is wrong and places our actions on a path to take corrective actions before whatever is wrong gets worse. Although the process of feeling pain can be an emotional one, the process of our body warning us about this pain is rather technical. When we are physically injured, the injury triggers nociceptors on the tips of our nerve endings. At the moment of injury, a message is sent to our spinal cord, then to our brainstems, and then to the thalamus, which provides our consciousness with the *perception* of pain and causes us to immediately react. This perception is connected to our sense of touch and emotion. All of this happens in a fraction of a second. The same reactive process happens quietly at the genetic and cellular levels. The process at the cellular level is technical and emotional at the same time. The signal pathway of the cell is responsive. It is instantaneous with no bias toward whether you give it positive or negative directions to follow. The cell responds to emotions; think heart rate variability or the emotions of love or rage. Both do very different things to the cell.

Until now, we have been limited by our understanding of what makes us who we are at the cellular level. What makes us able to do what we can do under the forceful hand of extreme pressure? Is it simply adrenaline rush or something much deeper? The celebrated exhibitions of man's ability are perceived anomalies, which is an incorrect analysis. Exhibitions of great courage, strength, and performance more accurately represent our true self. Our true self sends a flare into the sky of our consciousness, hoping we will follow its trail down to our cells to find out where all this physical and intellectual power came from. Return with me to that scene from *The Grey* once more. Liam Neeson is still kneeling next to his colleague—eyes locked, hearts touching. The authenticity of Neeson's concern has erased the line between personal experience and empathy. His friend's suffering is now his own. His friend's body is fully engaged in critical injury, but the weight is lighter. The pain is more bearable as Neeson has volunteered to bear some of the load—an authentic partnership. Important to assisting my clients in reaching peak performance, regaining their health, or avoiding sickness altogether is the element of partnership. Authentic partnership blurs the lines between "their problem" and "my problem." When the medical professional reduces their position as objective clinician and incorporates a touch of humanity, things change pretty quickly. What changes mostly is perspective, and often that's all you need—the right perspective on your problem.

Patient as Partner

Unity between patient and doctor improves healthcare. All parties must have a vested interest in a successful outcome. It is humbling how many millions of people spend the most vibrant time of their life working their bodies into the ground preparing to "enjoy life later." Later never comes. Sickness often creeps in right when retirement begins. Funny how that happens. Many find themselves unable to travel, run, enjoy family, and explore new opportunities due to a decline in health at precisely the wrong time. This doesn't have to be. Again, this does not have to be, but living an active life requires a holistic approach to hold on to the energy you enjoyed in your youth.

Many of our discussed ideas—such as hope, motivation, and our genomic triggers—and how these subjects relate to influencing the cellular structure for regenerative purposes counter conventional medical treatments. Conventional medicine concerns itself primarily with the naming of a disease in an effort to treat patients scientifically. Assessing the spiritual, biological, psychological, or emotional needs is simply not priority. Rarely are the environmental effects of toxins and medications considered, discussed, or shared within the sacred space of an examination room. Since all of these elements, the emotional, the psychological, and the physiological form the core of our human existence, how could one adequately treat human illness or promote health without taking these elements into thoughtful consideration? When the clinician stops looking at a sickness as a patient problem and instead makes the problems his own, things change.

Dr. Eller

In a small Pennsylvania town in 1968, a young woman sits still as her doctor speaks. Slowly, her right hand squeezes the corner of her bland hospital robe.

"What exactly are you saying?" she asks oncologist Dr. Eller.

He responds. "I'm saying the series of treatments we ordered have not been as effective as we hoped they would be. We will need to engage another course of action."

"Are there other options?" the young mother asks. "Is there an alternative approach that might give me a fighting chance? I'm willing to take risks because the risk of doing the same seems higher. Feels like I'm, well, feels like I'm running out of runway."

Dr. Eller seems to emotionally resonate but clinically disagree with the notion of a nontraditional treatment path. With his emotions under lock and key, he answers.

"We have proven treatments we must rely on to provide the highest probability of regaining your health. I've reviewed your case with my partners and we are in solid agreement with the new course of action." Dr. Eller prepares to exit the examining room when her words cause him to pause.

"Doctor," she says in broken whisper, "I need help. I need help from any source, any person, anywhere. I don't plan to sit here month after month fading away. My hair is gone, my eyes are dim and my skin bruises to the touch. If there is another option for my condition, on or off the record, please tell me not as your patient; please tell me as a mother of three children, a sister, and a friend to my church family. Please tell me all you know, even those things you're not supposed to tell me you know. Doctor, if you and I were joined in Siamese suffering, what would you do to make sure I would survive so you would survive? If my weight became your dead weight, what level would you go to for me to make it?"

Dr. Eller doesn't say a word. His humanity fights his training as he wants to respond from his heart and not just from his mind. He takes a break. He leaves her and goes into the break room to be alone for a moment. Her words were all too real. They jog his memory.

He remembers the day his brother donated his kidney so Dr. Eller could live. You see, this was long before medical school, before the money, before enjoying perks the title of "physician" can bring. His parents, who were immigrants like so many in Brooklyn during the depression, fought hard to make a better life for their boys, but without money or any real opportunities, they had to depend on the kindness of others to survive. They had to depend on the empathy of those who knew what it was like to have your life in the hands of another. Although just an adolescent, he remembered his mother telling him the story about the day they found out he needed a kidney. He remembered his mother crying a cry a child should never hear. His mother knew that her child living with such a desperate health care need meant he wouldn't live. But here he stands today, a physician in charge of offering patients a way to make it out of their sickness.

So the young lady's question is a heavy reminder of that day many decades prior. The day his mom told him his older brother would give up one of his kidney so he could survive. His brother risked his life and gave him a chance at one. And when David Eller was fifteen, his older brother died. David never really got a chance to properly thank his brother for what he'd done. So right here, under a bright florescent light, alone in a sterile hospital break room, Dr. Eller closes his eyes and utters three words toward heaven.

"Thank you, brother."

Dr. Eller walks back into the examination room. The young woman senses something has changed.

"Doctor? Are you ok?" she asks.

"I am."

He leans back onto the sink counter and uncrosses his arms. He seems to be fighting his thoughts before answering her question. He went to medical school because he wanted to help people, but his clinic is in the same crisis as all others. There are limited options for treatment. He feels restrained by the accepted, approved forms of care. Today wasn't actually the first time he thought about the limits of his patient approach. Several months earlier, his colleagues invited him to a functional medicine conference where he, for the first time, learned about the power of hope. For him, hope was something emotional, something sappy. Hope was something you relied on after science didn't work. But at the conference he learned hope was a science, hope was a strategy. Hope *is* a treatment.

At the conference, Dr. Eller learned of healing foods, studied more closely the complicated gene-biochemistry-performance connection and how these matters interplay with environmental factors. He went to the conference depressed but came home encouraged about the new possibilities available for his patients. In other words, Dr. Eller actually has the answers the young lady is seeking. He has a better way. The question is will he leverage the new treatment path on her behalf? Standing now in front of her, he speaks.

"For you, there is another option."

Dr. Eller goes on to detail the peripheral causations of her central illness, a list that includes healing foods, support, motivation, emotions, and the power of breath. He goes beyond "you are sick" into "why you are sick." From that point on, they engage in mutual discovery and treatment, as if her weight will become his dead weight if she doesn't survive. Dr. Eller accepts her hypothetical scenario. He starts the intake process over from scratch creating an extensive questionnaire concerning her chronological, environmental, and dietary history, and her personal support system. This new approach gives her hope that there is an answer buried somewhere and together they will find it. This willingness to go beyond the call of duty will affect scores of patients for years to come. Nearly five years later, Eller opens a functional medical clinic serving hundreds with this new patient approach and it all began with a patient unwilling to accept her fate.

Functional Medicine

To be functional connotes dismissing aesthetics and concentrating on the integrity of operation. The way something functions also shines light on inherent differences between two things in comparison. One may say, "There are functional differences between a bicycle and an automobile," to display how, although transportation is the core objective of both, the automobile has an advantage. When you think of the medical professional you've appointed as custodian, as advisor over your health, does this person give you a functional advantage over sickness and disease?

Did you follow the life cycle of hope between Dr. Eller and his patient? He was given hope as a child, as a gift, and through this gift he survived sickness. Years later, his patient refused to let go of hope even against insurmountable odds. She transferred this hope to Dr. Eller, who was motivated by her courage to challenge his approach. This motivation brought back an unpaid debt to his brother in his memory. This unpaid debt became a catalyst causing him to embrace a bold new functional pathway towards treating his patient. Doctor and patient became shareholders in success.

Memory has a powerful, creative physiology that is greater than the more seductive path of giving up and giving in. Did you know your cells have memory? Did you know they remember what your body looked like, felt like, and performed like when you were the younger, healthier you? Did you know you can motivate your cells to regain the memorized health of your youth?

Hope is Best When Hope is a Habit

Harrison Okene. Take another person and place them in that sunken ship on that night under those circumstances and they may not make it an hour, much less three days. Panic and despair may have overtaken them, disabling their ability to think properly and thereby extend their survival time. Why do some people make it and some people don't? Why do some people have the courage to challenge authority like the young lady challenged Dr. Eller, while others shy away? The variety of outcomes is not based on the lack of hope. This may come as a surprise, but everyone has hope. Yes, everyone. Much like falling rain validates high clouds above, so does despair validate high hopes above, from which it has fallen, but we are biased to anecdotal tales detailing hope's victories, which seem to more often arrive by surprise in response to surprise events. The person who may not have survived Harrison Okene's ordeal would not meet an unfortunate fate because of their lack of hope, but rather their resignation to the same. Another patient sitting in front of Dr. Eller may have remained silent and accepted the answers given to her. Another patient may have surrendered. Another would have pushed through those heavy rains of discouragement and reached for higher hope that sparks motivation.

Hope surrendered is breeding ground for despair. We rightfully champion those like Harrison who find hope and miraculously survive, but there's an even more powerful use for hope. It is hope as a constant nutrient instead of an occasional blast. But how do we arrive at such a place, the place where our thinking, behavior and approach to daily activities normalize hope? How do we make the fascinating less fascinating because we experience great all the time? How do we make the supernatural natural? We do this by intense observation of what works and then systematizing what others view as random. We do this by making hope a habit.

Habitual Greatness

BJ Fogg is the founder of the Persuasive Tech lab at Stanford University and author of the Fogg Behavioral Model (FBM). The FBM was designed to quantify what causes human behavior to change. The model maintains that three elements must converge at the same moment for a certain behavior to occur—*Motivation, Ability, and Trigger*—and when one of these elements is missing, behavior simply does not change (see model below).

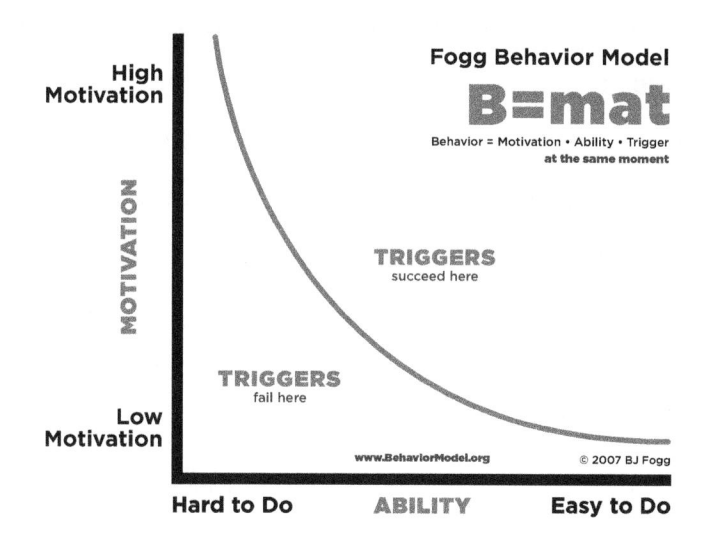

Fogg Behavior Model

B=mat

Behavior = Motivation • Ability • Trigger at the same moment

TRIGGERS succeed here

TRIGGERS fail here

High Motivation — Low Motivation (MOTIVATION axis)

Hard to Do — ABILITY — Easy to Do

www.BehaviorModel.org © 2007 BJ Fogg

The Fogg model was designed as a tool for individuals seeking consistency of performance. It is for academics and clinicians seeking to better understand the cause of peak or less than peak performance in their students, patients, and clients. The model provides unique insight into why certain predetermined targets or actions are or are not being achieved. In Fogg's thinking, moving great performance from the haphazard to the systematic increases the predictability of success. If you can trigger success, you can more often be successful. Fogg has identified certain core motivators holding the greatest influence over our actions. These core motivators are *pleasure, pain, hope, fear, social acceptance,* and *rejection,* and I would add one more: information.

One may argue that variables such as pain and pleasure are inherently informative, nudging our actions toward continuing one and discontinuing the other. But within the context of healthcare and wellness management, my patients value information over nearly everything else. Informing a patient there is a way out of their present sickness, an exit door leading back to the sunshine outside, changes everything. It is then paramount not only to have motivation to hope but to inform them of this hope so that the patient not only makes bold decisions but the right bold decisions. The courage to act is made into mockery if the incorrect action is taken.

This why the Hope Network—a structured network based on classical science and the science of hope—is so powerful in changing your entire health profile and bringing you back to strength, peak performance, adventure, and an overall greater quality of life. The network is triggered at the cellular level. We've discussed motivation and its power to turn the worst of circumstances into good. But few genetic triggers determine your life, its quality, and its length like this one thing: food.

CHAPTER FOUR
Rita

"She moved."

I could feel the unbelief on my face.

"She did?" I asked. The young lady nodded in affirmation. It was early and freezing cold, even by Pennsylvania standards. I'd sat down at this quaint coffee house not far from my office for an early morning cup of coffee. I didn't expect anything but a cup of coffee, but got much more. My server's name was Rita. I learned Rita was from a small nondescript town in Mexico just a few miles south of Acapulco. I took the opportunity to change the subject.

"Oh Acapulco," I said. "Heard it was nice there."

"That's what I hear," Rita said with the pleasantest of smiles.

"I've actually never been there."

"Oh."

Rita started making my coffee. The place seemed unusually quiet. I say this having never been there before. It just seemed different. The television was on but there was no sound. Through the port holes, I could see someone in the kitchen—white hat, head down. The young host was at the front on her phone, and one guy was sitting in the far corner reading a tattered book. I'd driven past this place for years but only when I took a different route to work, which wasn't often. On this morning I decided to stop. I looked up and caught Rita looking at me. Small frame, dark hair pulled back with a humble sentiment. It felt like she was thinking of saying something but contemplating whether she should or not.

"So how long you been working here?"

"Almost ten years."

"Like it?"

"It's ok, miss my family, but my husband and son are here."

"I understand. Do you go back and visit?"

She began steaming the espresso shots for my coffee and leaned back from the machine to answer.

"As much as I can."

She handed me the cup of coffee.

"Thank you, Rita."

Without transition and only a small pause she began to talk.

"You know I lost my sons."

The coffee mug hadn't yet reached my lip. I put it back down.

"My God," I said. "I'm sorry to hear that."

Rita's eyes began to sparkle. She wiped her hands on the damp towel tied to her waistband and her cheeks turned red.

"I felt like such a failure. Told my husband he was free to go. I mean, I couldn't give him the children he wanted. Think that's one of the reasons we left Mexico, too much pain there. Too many memories—just too many."

The man reading a book motioned from a distance.

"May I help you, sir?"

He wanted a power bar and his timing was perfect. The conversation needed some air.

"Yes, right over here."

Rita walked around the counter while I took a sip of my coffee as if it were liquor. She went into a cabinet and pulled out a power bar walking it over to the man. I nearly wanted to thank him for the interruption.

"Yes, told him he was free to leave. Didn't have any problems with him finding another woman. I couldn't bear the idea of him marrying me and then not being able to give him a family. Couldn't hold the pregnancy. It was my cervix."

"Incompetent cervix?" I asked.

"Yes! That is exactly what the doctor said it was. I had to find the right doctor though. Originally, I went to this one specialist. I can still see her hands moving like she was frustrated while she spoke to me and my husband. Now, I was already in unimaginable emotional pain when she comes out and says, 'Your body just can't have a baby, stop trying.' So discouraging. Couldn't believe she was a doctor. She gave me no hope, even worse; she took away the hope I had."

"Really?"

"Yes—that woman."

I took another sip, a bigger one this time.

"My first baby was a little girl," Rita said. "She was born right before thirty-one weeks. She was beautiful—looked like my grandmother. Long hair, long fingers, perfect in every way. I wanted them to do something. It seemed like they could have done something to save her. I mean I saw it."

"What did you see?" I asked.

"I saw her move. She moved."

My head dropped down to see my coffee spinning clockwise. I could feel the hair on the back of my neck standing up ever so slowly.

"You saw her move?"

"Yes, when Dana was born—her name was Dana—she quickly pulled her hands to her body like she was cold. Couldn't they have saved her? How did she know she was cold if she wasn't alive?"

"I don't know. I do not know Rita," I said.

Rita took the towel from around her waist and tapped the corner of her eyes. She looked around to see if her boss was catching wind of a personal conversation at work. She let off a big, yet quiet, breath, so I added a few thoughts.

"I understand; I do. The pain of losing a child is unbearable. Emotions can come out of absolutely nowhere and stay for as long as they wish. There is no routine, no decrease in hurt, no way out. Only a way in. A way into accepting what has happened. A way to live with the permanent limp of loss. When you lose someone dear, you never fully recover. I look at it this way, if happiness was ten before you lost a loved one, now it's an eight or nine. It never goes back to ten. You simply learn to be happy at a lower level of joy. That is how it works in reality, but few people are willing to talk about it, much less admit it."

"That makes so much sense," Rita said. My family has been for the last decade wanting me to get over it, saying how many people have lost children. It's the weirdest thing I've ever heard. There is a feeling like you have an unofficial amount of time to get over it and then, after that, something may be wrong with you."

A group of three walked in the door and tossed a command from across the room.

"We'll take one tea over at this table."

"Sure," Rita responded.

As she moved to get tea I began to think of what I'd just said. Words I wish had been said to me, to my mother, to my father. I'd grown up my entire life not understanding the advice I'd just given to a total stranger, a stranger who in some odd way had opened a door about something she needed to talk about not knowing it was the same thing I needed to talk about: my siblings. I'd felt a soft guilt all those years. I say soft because I'm appreciative of life, just wondering why I was allowed to have it. It's complicated, nearly beyond explanation, and only then, right there at the counter as Rita randomly launched that conversation, did her feelings free mine. She walked back over from delivering tea.

"People can be a little rude," she said in a low voice. "They kind of expect you to act like a servant when you're in the service industry," she added.

"So you deal with this often?"

"Every single day. One guy couple days back walked up to me and said, 'I need for you to make my eggs hotter, they're not hot enough!'" Rita kind of chuckled. "It just hit me as funny, never had someone say, 'Make my eggs hotter,' before. I hear everything."

"You must be a patient woman."

"I pray a lot. I believe in God, but you know what's even better? I've learned over the years that He believes in me."

"What a lovely thing to say, Rita, what a lovely thing to say."

"When I was dealing with losing Dana, I prayed, fasted, and had my church group interceding on my behalf around the clock. We did everything the Bible tells us to do. Believing is a tricky thing."

"What do you mean?"

"Well, God never allows your faithfulness to overcome His will.

No matter what we pray for, no matter what we believe, His *will* will be done. The tricky part is we don't know, we don't ever truly know if what we're praying for is what He wants. I keep using the word tricky because—well, there's probably a better English word, you know?"

"Yes, but I understand what you mean."

"We had pastors and people who've been in the church for fifty years all holding our little Dana up. We were hoping for life; after all, we were going to dedicate her back to God. She was His child. When you're right at thirty-two weeks, why would God allow my little girl not to make it then? He could have taken her much earlier. It's a mystery, and anyone who says it isn't just isn't being honest."

"I understand, Rita. It is a mystery how God decides to take one and leave another. Sometimes it seems the people who don't pray and the people who aren't good get all the blessings, their kids are well, and things seem to be going fine."

We both smiled knowing this was a violent oversimplification.

"What happened to your second child?" I asked.

"It was almost the same scenario, incompetent cervix, couldn't hold the pregnancy. We traveled to every doctor. They were mostly negative. Maybe they saw two young Mexicans and didn't think our children were worth saving."

"Don't say that!" I exclaimed.

"Well, some people feel that way, don't they?"

"I don't feel comfortable even answering that question. If they do, they have no association with me. Every child, every person, every single one is a gift from God without exception."

"Let me get you another cup. I've talked your coffee into being cold," Rita said.

As she turned around I started to think of my own experience and found sufficient encouragement in her boldness to speak about my own weaknesses.

"You know, when you were speaking, I couldn't help but think about my nine siblings," I said.

"Nine? Wow, what a big, beautiful family."

"They didn't make it."

Rita stopped the espresso machine and put the cup down.

"You've been so honest in telling your story, Rita, let me tell you mine."

CHAPTER FIVE
Food Force

Many years ago, my father came to a crossroad. To understand, you must join me next to him at the cemetery, burying his twins.

With one hand, Perry holds his wife, Lydia. With the other, he squeezes a tear-soaked handkerchief. With eyes closed, he takes one step toward the makeshift memorial made around the tiny cemetery plot. He grits and whispers, "Ode to Innocence Lost"—a creed memorized as a younger man.

> *After the curtain closes on the sun's last act, and the moon no longer lights the poor man's path, my child, I will love you still. I will love you more tomorrow than I love you today, in hopes that God, in His infinite mercy, sees fit to again send you my way.*

Standing there, Perry again embraces his wife. There are no words. Their children will never grow up. They turn to leave the twins behind, but even while walking away, Perry refuses to give up hope, the hope that one day Lydia will be able to carry a baby full term. One day, they will have a healthy family. To accomplish this, one question must be answered. The question is simply, "Why?" You see, this wasn't the first time Perry and Lydia experienced such soul-wrenching disappointment. All nine of their children died before birth. But why? How was this happening? What was going on in or outside of the home that prevented them from sustaining the lives of their offspring? They had exhausted every doctor. They looked high and low, soliciting advice from any and everybody they thought could provide insight into their grief. Nothing helped. And so Perry looked in the place where, in the end, most answers are found hiding. He looked in the mirror.

He was an engineer by training. Engineers tend to approach random occurrences with an eye toward locating patterns, even if the pattern exists in scintilla form. A keen eye can locate logic in the midst of illogic. A keen eye locates a *cause* while everyone complains about the effect. But the genius in engineering principals doesn't reside in interdisciplinary work, but rather transposing these ideas and then deploying this transposition into fields with no pre-existing defense against the new philosophical invader. We see this concept in medicine quite frequently. Quickly moving viruses are difficult to contain because of the virus' ability to morph, thereby avoiding defenses that were established based on their older version.

If one is to conquer health challenges like those faced by Perry and his wife, having the boldness to do what you've never done before is an imperative. The boldness to change your perspective can be the difference between life and death. How do you look at nature? How do you look at science? How do you look at your body and its ability to communicate what it needs to operate, to live, to thrive? Science needs nature, but nature doesn't need science. Nature doesn't solicit pharmacy for physiological redress. Nature exists in autonomy. Nature is self-healing, self-correcting. Consequently, in nature we find everything we need to eat properly and live vibrantly.

Perry is still there holding his wife. He begins to think about what is in their home. He began a process of reverse engineering to find systematic flaws in the family's living habits. *As parents, what were we consuming? Were we transferring toxins to our children? If so, where were these toxins coming from? Indeed, something was working, just working in the wrong way.* Perry and Lydia were experiencing a one hundred percent success rate of failure.

Immediately they changed their food. The exchanged cans, and readymade foods for fresh, colorful foods full of energy. They transformed their dinner plates into power plates. One year later they held their first, healthy born child. Was it the nutrients from the fresh food, a miracle, or the miracle of proper eating? What is dying in your life that could live if you gave it foods that were alive?

Engineering taught Perry that inside of every big problem is a small problem struggling to get out. He began looking for the small, solvable problem. He found it. In his thinking, if a child isn't born with a sickness, then it must have been given the sickness by some person, some food or some outside source. Sickness doesn't create itself. Sickness can be traced genetically, environmentally, communicably, or in many other ways. Sickness often relates back to missing one of the eight genomic triggers, like food. Perry found that he and his wife were unknowingly poisoning their unborn children. If it is processed food, it is poison. Perry knew that once he introduced a different nutritional regiment into his family, something had to change. While he could not predict what these changes would be, he knew one thing: *If every baby is dying, any change is going to be better than the status quo.*

Perry is my father. I, Ann-Louise Johnson, am one of the children who live today because my parents changed their eating. I am one of three of twelve. Nine of us died; three of us lived. It was 1946-1947 when my parents changed their food. My strong engineering-minded brother Perry was born one year later. Priscilla, my musician sister, was born in 1949. I was born in 1950. That is why I say we are 1:3:12. As a testimony of thankfulness, I spend my days helping others enjoy the adventures of life through proper eating and genomic care, but you gotta be alive first; you gotta be here to win!

The language must change. We don't eat food; we eat electrons. And it is this repositioning of our perspective that transforms our plate. Electrons, a derivative of electricity, are how we live. Arguably, one the most ingenious functions of the human body is that an organism made up of mostly water is fueled by electronic pulses from the heart. This is why heart defibrillators "shock" the heart to get it working again. The defibrillator is attempting to put into the heart what the heart already puts out—electricity. This electricity originates outside of the body. All electricity in nature originates from the sun. So when you eat, you are eating what the sun has grown by way of nature's electronic transfer. Think about what this implies.

It is not that we don't understand what to do; it is that we often think the body operates differently than the normal facilities around us. For example, we understand power, we understand energy, and we understand generators, but there remains a gap in application. It doesn't take much to fix this. A small thinking change can render enormous results as it relates to making quantum leaps in our health, but we first digest one fact: the body works just like everything else around you.

Each day, society effectively separates good power from bad power, or, better said, clean energy from dirty energy.

Environmentally, coal is the "dirtiest" source of energy creating the dreaded carbon pollution footprints. Conversely, wind is considered one of the cleanest. Now, both sources give us power, just different power. One source creates sustainability; another source creates more harm to the earth than it does good. But the same concept of sourcing clean energy to benefit our earth must be used to benefit our bodies. Body pollution, by way of diet, is equally, if not more damaging. The earth in us must be cared for as much as the earth around us. The question is how do we best do this?

The most important ingredient in everything we eat is its electron makeup. You see, food gives us energy because of the electrons found in it. Food is the car, but its electrons represent the engine. We must concentrate solely on the engine when we are deciding what to consume. This, in many ways, turns our concentration upside down. Traditionally, society's discussion around food has been based on, well, food. Conversely, we should actually view food as its most basic component, electrons, not its final makeup (cooked meals served at our table). This is a shift in thinking.

Think of food from the bottom up. Think of your next meal as selecting what electrons you will consume, not what carbs or proteins you will eat. That said, not all electrons are alive or "clean" to use our environmental metaphor. Not all electrons are good engines. But let's take one step back, back before the engine (electrons) and before the car (food). Before your food is harvested and sent to your local grocery, your food receives its power, its electrons, from the sun. Without becoming too technical, the electrons in your food come from the sun's UV rays that enter through the earth's atmosphere. The sun is the ultimate power source, and it places its power in everything underneath its rays, including your food. Nature is so beautiful, so vibrant and so eternal, but have you noticed there are no microwaves in the forest? Free roaming animals with thick manes and healthy teeth never open a plastic package or tear the sides of a cereal box to access their food. Nature feasts on pure electrons. The purest, cleanest form of food (energy) originates from digesting that which has been energized by the sun's energy and not modified, processed, or packaged by man for shelf life. Even man's best friend knows this.

Traditionally, a physician doesn't receive an education in nutrition in medical school. Ironically, that type of education is given to those who care for our dogs. Yes, a veterinarian has more education on nutrition than your medical doctor. This is why when you take your dog to see the vet, he knows immediately if you're feeding him the wrong food. He knows this because your vet looks at whether your dog's hair is falling out or other outward indications of an inward nutritional deficit. But the second reason most doctors don't concentrate on this level of healthcare is it takes time to listen to a client's story and really understand what's going on; doctors can't take this time while in insurance mode. It is fascinating to compare and contrast veterinary science and basic medicine. Veterinarians have more of an acute understanding of how electrons in food affect the body.

What makes this mismatch of vital information even more disturbing is this. Your dog cannot come into the office and speak, cannot communicate where he's hurting, and cannot communicate anywhere near human ability, so in a twist of irony, the veterinarian is taught to concentrate on what matters most. The veterinarian cannot be distracted by the communication of the patient in front of them. Whereas we, the intelligent beings, are generally allowed to tell the doctor what is wrong with us, the animal, who cannot communicate, many times gets better healthcare. Even further, think of prize cows, Angus, etc., we put so much breeding and healthy nutrition into. The only way a stallion can run at some of the big racetracks is to be fed pristinely and taken care of in the same manner. Look what happened in pet stores; almost overnight, corn-free, gluten-free, grain-free dog food became standard in every pet store around the country. Feeding your dog anything else is now is considered nearly abusive, yet humans still eat the same unhealthy foods we wouldn't feed to our dogs. It is then up to us to make up the difference.

So we see our food is first powered by electrons, which are first powered by the sun. We now understand that all energy is not equal. There is clean energy, and there is dirty energy. There is clean food and dirty food, and depending on which version of food we eat we will be faster or slower. We will be stronger or weaker. We will be alive or dead. You, and you alone, have the ability to select what role food will play in your life. It is, dare I say, our responsibility to give ourselves a fighting chance. Love requires we stay around for those we love, if only one more day. Wisdom requires we share our wisdom with those carrying the baton of our name. And it is high-grade electron foods that enable us to extend our time with our young and our days with our old.

Food is so widely misunderstood that most see it as a means of fueling their bodies for the day. We hear the phrase, "I need to grab a bite," all day long. Our busy lives have reduced food to an almost negligible habit done "on-the-go" to get us back to what we think really matters: working and making money. Obviously, this is a terrible deception. It's a deception because what we feed our bodies enables us to have the clean, sustainable energy to continue working. Remember our example: the engine first, the car second. We must begin to seek high-grade electron meals, not simply food for the sake of eating.

Another part of understanding what to eat and how to eat it is simplifying the language around food itself. We are constantly inundated with food acronyms that make us feel we are eating healthy. You've seen the signs—"No GMOs" and "No MSG" Most people have no idea what they mean or why they are avoiding them. Add to that the more important acronym, ATP (adenosine triphosphate), and even fewer people have any idea of how this affects their health. Let's discuss ATP for a moment in the light of understanding electrons.

Just like the money in your pocket, ATP is currency, energy currency. You see, as the food you eat is oxidized, energy is released. This released energy reforms into ATP so your cell can maintain a proper supply of the molecule. Suffice it to say, in your body right now, electrons are moving through something called the electron transport chain. When this happens, energy is released. Think of it as these electrons get tired just as you would if you were running back and forth. This energy is then used to add a phosphate to ADP (Adenosine diphosphate) to make ATP. As complex as this is, it is just as simple. The body, using ATP, stores, processes, creates, and consumes energy. The entire process centers around energy.

If one were to whittle down all medical jargon into one succinct phrase about human life, it would be this. To be alive and healthy is to be full of energy. To be dead or dying is representative of having little to no energy at all. Food is the outside energy that charges the inside energy source of ATP. The better the food the purer the electrons, and the better ATP can function. The better the gas the better the drive. The better the engine the better the ride. So then, using energy as our external and internal guide, it is relatively easy to identify foods that have energy and foods that have been robbed of the same (back to our clean and dirty comparison). Eating dead food, or eating worn out, dead electrons, kills you because food transfers into your body exactly what is in the food itself. Eating food that is alive promotes life for the same reason. For example, microwaved foods are extremely damaging to the human body. Microwaves must kill food in order to heat it as fast as you want it. For speed, you sacrifice life. Have you ever noticed how food tastes differently after being microwaved? To cook your food in seconds requires an electron pinging and cycling more than two billion times per second. The electrons may lose energy as a result. When this energy is lost, food tastes different. It is then no mystery that we may feel fatigued, tired, and lifeless in the afternoon after eating a microwaved lunch. Remember, food encourages us to be like it is.

We need foods that look and taste vibrant—full of life. We need healing smells and living colors of green, yellow, and red. We want to digest foods that are fully charged with electrons. Reject the lifeless plate in front of you. Stop eating dead food. Stop using the microwave. When we digest foods that are dead, our bodies must use ATP energy to process a substance that gives no energy back to it; therefore, the body is left in a deficit. This isn't a new concept. Just go outside and hook up your car to another car with a dead battery. Doing this once won't affect your car much. But if you spend an entire day "jumping" other cars with dead batteries, their dead cars will eventually make your battery just like them—dead. No person, place, thing, or ATP can constantly give away energy without eventually dying. We do all of this. We eat electrons to firm our physiological structure to ultimately resist sickness. Sickness is a hitchhiker. Don't give it a ride. Without our invitation, sickness has no home. We invite sickness into our bodies through the foods we eat, or, better said, through the energy-poor electrons we consume. Inviting sickness in is easy; evicting sickness is hard.

The human body is a well-defined, complex, superlatively powerful organism possessing infinite creativity, health, and life-giving functionality. The body's default position is to stay alive, and we want to stay alive. But to do this requires us to understand the most basic, fundamental reasons as to why a person stays alive. Why does one person live a long, healthy, vibrant life while another is pained by sickness all of their days? While there are unavoidable, genetic ailments many suffer from through no fault of their own, most diseases are self-inflicted. Given the chance, who wouldn't extend their days on Earth enabling us to hold our children's children in the bend of our arm? We want to be here. Who doesn't want the pleasure of seeing that sparkle in the eyes of the young woman our son is taking as his own? We want to be here. This is the true reason for healthy living and proper eating. We want to be here for those who matter most, if only one more day.

Statistics are bullies. Mortality rates are passive aggressive. Reject their diabolical claws. Numbers can never predict your lifespan. Only you have that power. Hope, on the other hand, is forceful benefactor, a kind dictator of sorts. Hope allows one person to live a life no one in his family ever lived. Hope refuses to give in to history, status quo, or low expectations. And so it is, ironically, more complicated to modify what we've always considered to be the "right way" than it is to do something totally new. The historical understanding of something or some process creates dedicated populations of people who reinforce its reality. Just look around. Planes look like birds, and cars look like bugs. Buildings were patterned after trees and pools after the ocean. At one point in time, there was great resistance to bringing these derivative creations into being. Thankfully, hope is an inexhaustible soldier forever fighting for progress, even if it must fight alone. One day, you will look at a plate of food and see vibrant or exhausted electrons. You will have the ability to identify the invisible makeup of the food simply by its appearance and know the process by which it was prepared. The fork is the deadliest weapon on Earth.

Food is Force

Like me, you may remember that high school physics course where we were first introduced to the power of force. When a force acts upon a fixed object, only three things can occur: proportional elongation, disproportional elongation, or ductility. Proportional elongation simply means a force will be met with an equal force. Disproportional elongation refers to the reactive force being greater than the initial force imposed on it, and ductility refers to the amount of force an object can withstand before fracturing. These potential consequences provide insight into the character of force. Force forces change. Food is our greatest force because food alone can change your life.

The force of requirement: Food is our greatest force because no one escapes seeking it.

The force of regularity: Food is our greatest force because not only must we seek it, but we must seek it all our lives.

The force of mutable purpose: Food is our greatest force because in it lies the power to comfort and heal or destroy and kill.

The good news is this: There is no better time to be alive than right now, right here, today! The level of understanding as it relates to how food intake affects not only our weight, but the efficient operation of the human body is at an all-time high. The implementation curve between research and practical changes in growing, cooking, and eating has been shortened. This can be seen in the innovative area of *genomic kitchens*. Genomic kitchens are a part of a healthy eating and drinking movement called *Field to Plate*. This movement was made popular by Amanda Archibald who has successfully exposed new pathways from the kitchen to our molecules within. Archibald has highlighted why certain foods are molecularly more important than others. Your genomic kitchen is the "kitchen" inside of you able to work with the bacteria in you to create the health you need. In other words, your genomic kitchen represents what you eat and store in your gut on a daily basis. It is your food pantry. Is your food pantry full of sugar, salt, and bread or is it full of high-grade electrons, food full of the sun? Your kitchen can affect far more than just you. Ask my father.

☙

There can be great difficulties when trying to make substantive changes to our daily food intake. I get it. Oddly enough, even when all the benefits of living longer and stronger have been articulated, people are still slow to change. The problem is premise. In order for proper food to take its proper place in our diet, we must recalibrate our definition of what food is in the first place. We've discussed food and its relationship to ATP. We've discussed how food is the outside energy source fueling our body's inside energy source called ATP, but food still suffers from mass misunderstanding. We've all heard the phrase, "You are what you eat." This is partially untrue. Yes, food wants to change our bodies into what is in the food itself. This is why eating dead food kills you and eating food that's alive helps to keep you alive. Still, this cliché assumes too many things. It assumes the body is a passive organ seeking definition when, in reality, the body is already well-defined. Food has a very different place than what first meets the eye.

The earth has all it needs to survive and flourish; after all, the earth was here long before its inhabitants. The presence of inhabitants can benefit or hurt the earth. Irresponsible carbon foot printing, uncontrolled waste, and increased pollution in the beautiful oceans of the world add to the degradation of what was beautiful and powerful before we got here. When companies choose to build under efficient LEED specifications, recycle, or just take on a more responsible ecological stance, the earth benefits. However, the earth is not what we make it; the earth does not seek identity. It already has one.

Like the earth, the human body is autonomous in nature.

The body is magnificent in its power of computation, superior to all other species in intellect, creative power, and ability to procreate that which is stronger, smarter, and more intuitive than the previous generation. Food elevates the pre-existing genius of the human body. Food doesn't make us; we use food to continue what has already been made. It is then more important to understand the body before we understand the food we put in it. What helps the body perform at optimum level?

I have been called upon to work with Olympic athletes who, by any measurement, perform at levels higher than any person alive. Still there's more. Using this very approach of identifying their genomic footprint, using this footprint to determine the foods best for their body type, combined with incorporating ATP maximization and cellular rewrite, we've been able to increase an athlete's performance by as much as twenty percent or more (self-reported). You may not be an athlete; you may be suffering from diabetes, high blood pressure, or a number of ailments. Your only interest may be, "What increases or decreases lifespan?" The first step toward eating better is to take the simple step today toward eliminating all dead foods. Strip all food from your kitchen that has been stripped of its nutritional value. As a start, eliminate the following items from your diet.

Aspartame

A commonly used artificial sweetener, the amino acids in aspartame literally attack your cells, even crossing the blood-brain barrier to attack your brain cells, creating a toxic cellular overstimulation called excitotoxicity. Excitotoxicity refers to the ability of glutamate or related excitatory amino acids to mediate the death of central neurons under certain conditions, for example, after intense exposure.[5] Why should you care? Excitatory amino acids are neurotransmitters and in some cases they act as neurotoxins. They have a dual purpose. Point is you don't want your neurons to die. Your neurons are the key players in your brain. They are at work right now as you read this. Neurons are information messengers. By way of example, imagine driving down the highway for 500 miles toward a city you've never visited, but there's a problem with your trip. There are no signs, no stop lights, no nighttime highway lighting, no names of states or cities on billboards, and no exit numbers. How would you know where you were going? You wouldn't. Now add neurons to your trip. Everything lights up. Through electrical impulses, your brain's neurons give you information and direction. Neurons are the conduit between your brain and the rest of your nervous system. It is impossible

---

[5] https://physics.stackexchange.com/questions/67970/surviving-under-water-in-air-bubble#comment137400_68008

to read, think, or feel without them. They tell you where to go and how to get there.

Now back to the aspartame artificial sweetener sitting on the table at your local diner. Many of your brain's neurons will die naturally. Aspartame increases this natural degeneration and elevates the probability your neurons will die quicker. All of this in the name of low a calorie diet. Diets that suggest you ingest aspartame are properly named as the root word describes what's actually happening to you. Aspartame kills—a synthetic death only equaled by the dangers of ingesting pure white sugar.

Preserving our money, cars, homes, and clothes is a great practice. Preserving our food for the purpose of eating it months or years from now is toxic. Eating healthy is fundamentally centered around eating fresh, time-sensitive food properly prepared and void of pesticides or any other man-made altering chemicals. The only purpose for them is to increase manufacturers' profit. Unfortunately, chemicals used to extend the shelf life of food include hard to pronounce substances like butylated hydroxyanisole (BHA) and butylated hydroxytoluene (BHT), both known to adversely affect the neurological function of the brain. If you're discerning a theme so far it is this: our foods and the toxins found therein are designed to directly affect the cognitive seat of the human body—the brain. The chemical preservative tertiary butylhydroquinone (TBHQ) is so deadly that only five grams will kill you. Why do manufactures love TBHQ so much? For one, it cuts down on food odor so that if your food is rotting in the box you won't smell it when you open it. TBHQ can be found in everything from chicken nuggets to cookies and cereals on the shelves of your local grocery store. Do you want this in your body?

Synthetic Trans Fats

Fred Kummerow is an emeritus professor of comparative biosciences at the University of Illinois. Born in 1914, Kummerow has spent over seventy years studying heart disease and the causes of this disease. In 2013, he filed a lawsuit against the FDA after it ignored his petition calling for a ban on artificial trans fats, stating "Artificial trans fat is a poisonous and deleterious substance, and the FDA has acknowledged the danger." On June 16, 2015, the Food and Drug Administration moved to eliminate artificial trans fats from the US food supply, giving manufacturers a deadline of three years. Kummerow was pleased. "Science won out," he said in an interview. "It's very important that we don't have this in our diet."[6] But, we ate these trans fats for over half a century. Although we can stop now, they are in us and will continue to make negative changes in our bodies for years to come. We must

---

6 https://en.wikipedia.org/wiki/Fred_Kummerow

deal with and overcome the unhealthy inertia of our past.

White sugar, synthetic sweetener (aspartame), and synthetic trans fats make intensely negative changes in your body's functionality. When we talk about the epigenomic change, we refer to changes in your body that do not necessarily originate from your underlying DNA sequence but rather from add-ons. Add-ons refer to nutritional eating habits that can change your gene expression. This expression we speak of refers to how your genes interact with the world in vastly different ways. And these different ways are based on the quality of agitation they receive. Your nutritional intake, your food, is the agitator. Your food is the body's director and conductor.

When I think of conductors I think of orchestras. What a sight to see so many musicians on stage playing as one synchronized unit. At times, it baffles the mind how different instruments with different sounds like horns, percussion, and strings can be in such harmony. Even their movements are solidly locked. Arms move up and down in sync. No one is out of place—that is unless the conductor is. I've learned that the quality of the conductor has everything to do with what we hear as a listening audience. I've learned that if you take the same orchestra and place before them three different conductors, you will hear three different orchestras. Same players, different leader. This is because while the players are taught to play the notes on the page, they are trained to express these notes based on the instructions given to them by the conductor. Playing slowly, fast, quiet, or loud is controlled by the conductor. Your body is no different. The eight genomic triggers listed in this book provide direction. Each provides your genes with commands for how they will express fully or partially in your life. So when a person has a poor support system, their "isolation genes" become more pronounced, which changes their behavior, which changes how they interact with people around them. Even the expression on their face often changes. Everything touches everything. Everything works together for our good or bad.

So now you can see how when a person has a gene propensity toward a certain disease, the quality of her nutritional intake can increase or decrease the full expression of this disease. Her nutritional habits can be the difference between her bias toward a certain sickness or the actual manifestation of this disease. Her nutritional intake can suppress or amplify. Individuals who've grown up in alcoholic environments may have a propensity towards alcohol abuse. Persons in this position sometimes avoid alcohol for this very reason. Their threshold for addiction is much higher because of the environment that shaped them. Through their alcohol avoidance, they reduce or prevent the full "gene expression" of what could have grown into a major addiction. While there are admittedly differences between this example and, say, an individual born with a genetic, intrinsic ailment, there are more similarities than we first we see. When you fully grasp the rewriting ability of our body's cellular framework and how our genomic triggers represent the pencil with which we write our future, everything changes. A person's psychology is understood to have more to do with their success than their pedigree, education, or skill. Their thinking, their attitude, is what increases or decreases the effectiveness or ineffectiveness of everything else they possess. Their attitude is the trigger, so much so that even a person void of a great education, skill, or ability can very often reach the height of success because of their thoughts on life. Their attitude and perspective were able to rewrite the requirements, and thereby grant them access to a level for which they otherwise do not qualify.

Genomic triggers can do for your cells what attitude does for a person's career. Triggers and the psychological components therein change the rules by expanding, contracting, increasing, or decreasing a function. Triggers can feed disease and grow it or starve disease and kill it. Triggers can create athletic performance greater than the athletic ability of the individual. Triggers can change you into the greatest you.

Your epigenome is influenced by what you are exposed to on a daily basis—food, toxins, toxic or healing emotions, sleep quality—aka the eight genomic triggers that can push or pull epigenomic instructions towards health or illness. Your life is a song your genes must play. How do you want it to be played? What do you want your life to sound like? A fine-tuned orchestra or disjointed chaos? Adventure and bold happiness or sluggish, mundane melancholy?

Decide today because life is fast. By the time you read this, its speed will have increased. This places those of us seeking to overhaul our nutritional intake in a very challenging position. There are no two ways about it; eating correctly takes time. Bad foods are easily prepared. Bad foods are packaged to extend their shelf life, not to extend yours. Often, the best dietary information is ignored by the masses because to execute the proposed advice is an unobtainable ideal. Most Americans don't have personal chefs able to mathematically balance each meal. Most mothers trying to get the children ready in the morning don't have the luxury of juicing patiently while looking for little Johnny's perpetually lost shoe. This is why a prudent approach to bettering your nutritional intake is nearly as important as the food itself. Any healthy practice that cannot be sustained—no matter how beneficial—is worthless. Do you have a practical process for healthy eating?

This is a tactical question. Food selection will always flow in the direction of least resistance. If one must grab a quick bite, usually the first thing available will suffice, whether it is healthy or not.

That is why starting at the easiest place for change—and using the easiest frequency for such a change—is the firmest foundation to maintain the change. This is why I teach the program *Heal with a Meal*.

Concentrating on one meal per day is doable for most people. Over time, this one meal will create a space in the day when you feel better than the rest of your day. This creates momentum inside of you, and what's inside controls everything outside. Only what is inside of us has the ability to overcome the onslaught of bad dietary suggestions promoted to our senses all day. The commercials are bright, the food shines under the lights, and the cheese pulling away from that bread looks heavenly. This is powerful information, information we must process whether we want to or not. We must retrain our mind and body with better information in order to make better decisions. We must change our associations.

Population of the Air

I often tell clients that there's a population of the air—a society without a census. On these individuals no stats exist. There are people who have the knowledge and willingness to help you and me get out of every health problem we now have. The challenge is they aren't on the ground. It is not until you jump that you will meet the population of the air. It is not until you take a nearly illogical risk that you will find those who are waiting for you in the air. In the air, you'll find people who think, live, and move differently than anyone you see on a day-to-day basis in the general population. Ironically, this concept is difficult to express to those who are on the ground. And when I say on the ground, I refer to "ground thinking." Jumping means you are seeking to engage a new perspective on those things old to you. A diamond found in an African cave by a novice can look like a dirty rock if you don't know what is in your hands. The food on your plate can look like just another meal if you don't know the value of what is in front of you. Value is everywhere, but you have to get your thinking off the ground in order to find it.

In earlier chapters, we discussed the multidimensional nature of hope, but many are unaware of the hope-like nature of food. Many don't see the value. Food from the earth has been here forever. Water, nuts, grains, and proteins are eternal. By inference, we can leverage what has been here since before we arrived to help us stay here longer. Man isn't our example—the earth is.

We more often associate food with pleasure than with rehabilitation or healing. This makes it hard to recognize the magic that's always been there. If we leveraged food properly we could eat our way out of sickness, eat our way into health, eat our way out of depression, and eat our way into happiness. This alone births hope. Hope is a bottomless pit of potential.

It is difficult to contain hope by way of definition because it becomes what it needs to be for the person it is helping. *Food is no different.* Hope is a revitalizing cell. This revitalization, this redefinition, isn't a Pollyanna thought of the naive, but rather valuable gems from the innovative mind committed to finding a way out of no way. It is innovative in the trend of DaVinci—seeing and creating what has never existed and, by sheer belief, forcing it to now exist. And so it is, ironically, more complicated to modify what we've always considered to be the "right way" than it is to do something totally new. This presses the need for us to view our plate, our stomach, and our health in new ways. All that we need, like that diamond, is already in our hands, but we must view our hands in a different way. All you have is exactly what you need. If you want to live differently, think differently.

We started this chapter discussing the forceful nature of food. Force, by classic definition, is an influence tending to change the motion of a body or produce motion or stress in a stationary body. The magnitude of such an influence is often calculated by multiplying the mass of the body by its acceleration, a person or thing regarded as exerting power or influence. So, by saying food is force, we are saying that we can "force" our body to move in the direction of our choosing by choosing the right foods to put in it—and that is exactly what we mean to say.

Food is information. And like any other form of information, the better it is the higher the quality of data and the more likely you are to make better decisions. Bad information almost always leads to bad decisions. Do you remember the questions asked in the introduction of this book? Do I have a chance? Is there a shot? What do I need to do? These three questions define structure, function, and behavior. These three questions take us from the helicopter view of our problem all the way to the ground where the fight really takes place. Structure gives us the geography of our problem, function provides the mechanics, and behavior gives us the topography or the practical view of our challenge face-to-face. In our introduction, the questioner was seeking information. His inquiries were directed toward doctors seeking to manage the expectations of a Hall of Fame baseball player looking to somehow revive a career by reviving his body, a body no longer able to perform.

The importance of these questions must be understood. Hidden questions lead to hidden answers. Ambition to find positive people uncovers positive people looking for you. An unwillingness to accept sickness and the pessimistic statistics that accompany it leads you to others operating in the same sentiment. This alone can cultivate your healing through collaboration, motivation, and support. It again bears mentioning that there are no rules hope must obey; rather, hope writes the rules your body must obey. Hope and food are both forces. Newton's third law explains this nearly by mistake. All force submits to greater force; hope nourished by food is a force to be reckoned with. Hope ignores statistics and pays no attention to reality and facts. Hope ignores medicinal science. Hope is self-centered. Hope will not settle. Food is similar in many ways. Food is aggressive. While neutral inasmuch as it has no inherent preference for your health or sickness, food will take a pathway and not relinquish this pathway until its job is done. While dead foods will not be satisfied until you are dead, foods brimming in bright colors and nutrients will not be satisfied until you glow and are nutritionally sound.

So think of food in a new way, not in the way of recipes or comfort food, not breakfast/lunch/supper, not fats, carbohydrates, or proteins. Think of food as being dead or alive. Think of food as the conductor of your life. Give your gut what it needs to give back to you what you need. Nature lives forever. Eat what nature eats and you'll find your life extended beyond statistics, predictive family history, and mortality rates. You can rewrite your story today, one fork at a time. Pick up your fork and write your masterpiece. Your life is a song.

CHAPTER SIX
Sleep

To study sleep is a course in miracles. While sleeping, the organized complications of your cellular, circulatory, respiratory, and muscular systems operate without conscious command. Only in the sleep state does your body enjoy comprehensive access to its beauty, performance, and intrinsic genius. In the sleep state, your body heals itself and refuels your immunity to face dangers awaiting you when you awake.

Without rest, health is impossible. Yet, we have an odd, contradictory relationship with the practice of rest. On one hand, the successful, ambitious professional is one who works around the clock with little rest. She is in constant motion, which grants her the "go getter" classification. Modernity offers a crippling narrative—convincing the working public that sleep is an interchangeable act not enjoyed by the successful; that somehow working around the clock is a laudable practice showing strength and commitment to something more important than our health. Whatever could be more important than our health remains to be discovered. Nothing about perpetual activity extends the body's creativity or life. In fact, the opposite is true. It is when we nearly burn out that we tend to engage proper perspective on rest. We break for a little "rest and relaxation," which is precisely what the body needs as our daily reality instead of an escape from reality.

The human body is magical because it can operate at high levels of performance for long periods of time, without optimum conditions. Without optimum conditions, the body can be described as having a system of redundancy much like aeronautical design for commercial airlines. And like a commercial airline suffering from the loss of hydraulics, redundancy allows the body, wounded from life and hard living, to remain functional. This is how millions of hardworking, dedicated Americans in generations past spent thirty plus years on graveyard shifts in unsanitary factories or in coal mine country. In such environments, their bodies fought every imaginable toxin, but they kept going. We have met salespeople who cover the country selling widgets just to feed their families, all while smoking three packs of cigarettes a day, drinking perpetual coffee, and eating fast food with no exercise. The body keeps going. The body is resilient. The body wants to stay alive.

But have you ever noticed how many of these individuals, after living such a life, take on sickness, and die almost as soon as they enter retirement? It is as if the body takes advantage of the first opportunity to rest and places itself in the ground out of fear it may never have the chance to rest again.

Today, we are more educated on wellness. An octopus of wires connects patients to monitors in sleep research centers across the country. We know so much, still we sleep so little. Many of us don't properly care for our emotional health, which adversely affects our sleep. Many remain in toxic relationships and, despite these liabilities, the body still performs. So how much more could the body do if conditions that were favorable to its peak operation were employed? Few of us know. However, as clinicians, we know this: the highest priority enabling favorable function centers around rest. We must sleep.

We must sleep because fatigue is the father of bad decisions. Whether tired physically or emotionally, it is a proven principle against regret—we should never make important decisions without rest. This principle competes with a society that disagrees. Our culture praises "busy work." Our culture praises movement and activity, only because we don't understand how rest is more productive, more profitable, and more sustainable over the long term. It is ironic, but we consider one to have attained peace only when they die—thus we offer the tardy suggestion, "Rest in Peace." Why the morbid connection? Maybe because few seem to accomplish living in peace, which makes interesting questions arise. Have you ever paused to consider the connection between death and rest? Better said, have you considered the connection between death and sleep? Have you first considered what it really means to die? Dying isn't the opposite of living; dying is the opposite of sleeping.

Sleep is vibrant, resilient biology in full form. Sleep is an action verb. During a normal business day, your body is in a fight. Not a physical confrontation, but something much worse. Your body fights cancer cells all day. Without your knowledge, counsel or permission, your body fights bacteria, fungi, and viruses on your behalf. Your body is a soldier and a commanding general, keeping biological enemies at bay. And like a soldier, your body needs sleep to keep fighting, but not just any sleep—sleep at the cellular level. The cell is where everything lives or everything dies. Your cell is your seed and from it grows your life.

Cell as Seed

As an agricultural necessity, seeds are the center of life and the life birthed from its internal potential. The fractal characteristics of seeds—or their infinitely repetitive yet strengthening geometry—allow us to confidently plant acres of orange seeds. We plant these seeds knowing orange trees will emerge from the ground each and every time. The recursive traits of seeds can seduce us into believing what grows from the ground represents the full potential of what is in the ground. This makes it difficult to digest that most seeds wildly underperform.

When speaking about the sacred hand of nature, underperformance is a strong, even irreverent word; still, the description is accurate. The overall-wearing farmer understands this better than most. The farmer understands the seed's power, its generational influence, and, ultimately, the connection between cultivation and harvest. He also understands the fragile relationship between what he can affect and what is beyond his nurturing reach. To reach full potential, seeds must not only be fed properly, but they must be seasonally-planted to avoid fighting what naturally works against them and instead embrace what naturally works on their behalf. And what works most powerfully on behalf of the seed is dormancy. Yes, seeds sleep.

In each cell we find not just fractal powers of reproduction but blank canvases to create new forms, new pathways, new health, and extreme performance never witnessed. Like the seeds in your garden, the cell must be understood before we can leverage all it can be, thereby determining all we will be. Most perspectives of the cell are third-party in nature. Scientists explain what it is, where it is, and estimate how many there are. But what if the cell itself could speak? What would it say? This question seems to have only a theoretical attraction. Certainly, the cell cannot speak. The cell has no communication abilities, at least not any able to be heard by humans, or does it?

At a bioethics conference in Atlanta, Georgia, I was privy to a rather intriguing dialogue on the subject of cell-speak.

"The cell communicates," my friend said. "We just have to learn its language."

The friend I speak of is James Porter. Dr. Porter is a university professor and a staunch advocate for futuristic thinking in the scientific community. Science often has a biased objective to prove or disprove a theory, while James is more interested in finding ways to explain those things science has given up on or areas science dare not even consider. It is important to note, James is also an atheist, but an atheist of a different class, which I will explain in a moment. So we are sitting at a dinner table surrounded by friends, most of whom are medical professionals. These dinners have been known to erupt into heated debates on various topics from politics to medicine, ethics to capitalism. On this night we were dining at Sotto Sotto, an Italian restaurant near The Carter Presidential Center. The topic of interest on this evening centered around disruptive discoverers in the human cell. With his arms extended, James continued his treatise.

"How many at this table speak fluent Russian?" No one raised their hands.

"So then, does it follow that Russia doesn't have any great communicators or that Russians are unable to express themselves in any lucid manner? Of course not. You just don't speak their language. The body speaks, but until we stop thinking it must speak in American English or, better yet, until we stop thinking it must speak in scientific terminologies we understand, we will always lag behind the wisdom it offers in its own language. We often don't hear the body's advice on the front end; we hear its words in the form of physical consequence. Consequence seems to be a language we understand. Isn't that what an autopsy is? Isn't an autopsy the language of consequence? That final exploration infers backward until we arrive at causation for the patient's demise. But health also speaks. In fact, health and life speak much louder than sickness and death."

"Who ordered the chicken scaloppini?"

James looked at the waiter with absolute disdain and rolled his eyes. It should also be noted that James is ultra sensitive to interruptions and the chunks of meaning that interruptions can take away from an argument's momentum. The waiter set the plates down. James waited for calm to return, took an audible breath, and continued with one index finger pointing.

"Everything about our physiology is based on one thing—the cell."

Yoni, the cardiologist and wannabe comedian, chimed in.

"So true. I can't live without mine."

"Not talking about your iPhone, Yoni!"

Yoni smiled and winked at me. He always told a joke whenever James got too serious.

"Ok, ok—have a heart," the cardiologist said.

The table burst into laughter, all except James, which made it even funnier. We settled down and encouraged our friend to continue.

"As I was saying, the cell is everything. And while we speak and teach, and lecture about it, have we ever allowed it to speak, teach and lecture us about itself?" James took a quick bite of chicken, moved it to one side of his mouth and kept going.

"What is it that makes us who we are? What is the process by which the genius of the human body becomes the human body? Do you remember the story of creation in the book of Genesis?" Everyone dropped their silverware. We were all thinking the exact same thing.

"Did James just reference the Bible? Did the world just end?" And this is what I mean by an atheist of a different class. He believes there are certain narratives that explain great truths even though the narrative itself is false. In other words, although James doesn't believe the Bible to be true, he believes its stories are wonderful analogies to point to things that are true. That said, using the Bible as a reference point for his argument was noticeably odd. Using air quotes, he continued.

"What did God do when he decided to create woman, the

greatest species on Earth? He put Adam to sleep. Now just think about this for a moment. I don't have to recite my views as an atheist; you all know them. But this isn't about that, this is about conceptual truth that has proven itself to be valid scientifically. Sleep is the greatest creator, greatest grower, greatest cultivator of life. After conception, what do we do? We sleep in the womb. We remain dormant for nine months. Once born, all we do is sleep. We spend half our lives asleep. We grow older, and as our sleep continually increases, one day we go to sleep and don't wake up."

This seemed to resonate with the table. Everyone was eating slower, chewing their food in a noticeably thoughtful manner. Sleep is the most important activity of life. Sleep. To hear these words from James placed me in a very difficult position. The contradiction was obvious. Yet it made sense, and the more I thought about it the more sense it made. An atheist had just used the Christian Bible to make a point about the importance of sleep.

"Does anyone else see the problem with this?" I asked the table.

The table ignored me. No one seemed to. James, an intellectual powerhouse who never shied away from leveraging counterfactual angles to shock the listening mind into seeing a more important point, decided on this methodology today. In his mind, the best truths are built on lies. He continued.

"The cell can speak if we take time to learn its language, but even if we did not have the benefit of learning its language, we can still understand all it has to say."

Another friend at the table who'd been silent all this time jumped into the conversation. "I know what he would say," Dr. Newman interjected.

"Oh, do you now?" I asked.

"In fact, I do."

"And the cell is a 'he'?"

"He is," Dr. Newman answered without blinking.

Dr. Newman and Dr. Porter are cut from the same cloth, both very passionate about their views on the human body. Newman is a research scientist for a global biotech pharmaceutical corporation, and stem cells are his specialty.

Dr. Newman continued and placed his narrative around Dr. Porter's setup.

"Let's call this cell Winston."

Dr. Newman took a swallow of water as if preparing for a marathon.

"This is good, really good!" Dr. Porter said with a touch of sarcasm.

"I wanna know what little Winston has to say. How will he instruct and educate us concerning what makes the cell perform at his highest potential, which in return makes us perform at ours?" And with one open hand slap on the table, Dr. Porter turned the dialogue over to the biological voice of Winston. No one expected what they were about to hear.

☙

"My name is Winston. I knew you before you knew yourself. Who I am as a cell must begin with your concept of time. Time, a creation unique to the impatience of man, limits the degree to which you can understand me and my biological colleagues. Cells are offspring of eternity. Eternity stands over time as you stand over a caterpillar, seeing its beginning and ending in a glance. Because of this perspective, cells exist before your birth and continue to exist long after you die. This you must grasp. Circadian rhythms, or those twenty-four-hour rotations you have observed in plants, animals, fungi, and cyanobacteria, appear in our daily activity, but for us this is downgrade. We must abridge our true selves so you can somewhat understand what we do. That is, the cell naturally operates on a much more complex rotation—one you currently have not the technology or education to observe much less grasp in meaning. You've also noticed circadian rhythms to revolve around light and darkness, but the cell can also rotate around motions and emotions, along with a host of other variables. Understand this: everything you know is the least of what can be known. Cells are social, touching neighboring cells all day. Cells remain alive long after you die with a dog tag-like ability to point backward to the identity of the host. The world of the cell is much larger than yours, our intellect greater. We are slower than history. We are faster than the future. Often, you discuss matters of predictable biological events based on man-made clocks, but never forget this has no bearing on the operation of cells. Cells can operate within the confines of your timing, but we do not need your timing to operate. As eternal beings we conduct the body's business based on distance, not time."

Winston continued, "For example, in your world the smallest measurement of time is Planck time. You say this represents the time it takes for light to travel. Your theoretical physicists believe this to be the smallest measurable unit of time possible. In fact, there are more than a hundred smaller gradations of measurement familiar to the cell. Humans are at least fifty years away from discovering this. Notice something: when you discuss the speed of light, you do so based on distance traveled first and time expired second. Distance, for the cell, is unquantified, nonlinear biological activity without the framing of time exhaustion. The angle of eternity allows the cell open architecture to limitlessly solve, create, grow, and execute. Distance is a greater indicator of your personal accomplishment without artificial comparison as to how long your accomplishment should have taken. One may say, 'How do you quantify distance outside the context of time?' I'm glad you asked. Think of it this way. How long will it take the earth to run out of time and die? For humans, this is an unanswerable question because, for one, the earth isn't paying attention to time, you are, and second, without the earth paying attention to time there is no debt, accumulation, or compounding fatigue it must overcome. There is no approaching mortality rate in which the earth believes. There is no liability or disease that will eventually overcome the earth and kill it. The earth does not die because its existence isn't based on artificial construct. The earth isn't aware of statistics detailing how long a normal earth should live. The earth has no peers. In many ways, the creation of the time continuum, the very knowledge and *belief* in time, has embedded in it an intrinsic demise.

"For the human mind, data, statistics, and the mortality rates I mentioned, create a morbid synchronization based on artificial artifact. Whether atomic, sun dial, or the ancient Egyptian practice of clepsydra-based water timing, the issue is not what tool is used for keeping time but the assumption that there is time to begin with. Further, it is rather presumptive to believe your biology, which existed long before the creation of math or science, follows such a system—a system the details of which were man-made, man named, with definitions and connotations based on ideals man cognitively understands.

If a minute were 120 seconds, would you grow older faster because one weekday would now be forty-eight hours in length? If so, would you live longer if that same minute were now thirty seconds? What you believe time represents, creates what you believe it represents. If you believed time represented something else, it would represent that. This is why belief is the key—a dangerous key. Belief creates everything, life, and death.

"Cells are flexibility. If cells could speak to you as frankly as I speak now, you would seek cures and treatments exclusively based on an analysis of the individual without consideration of historical information on their sickness. This is counterintuitive as scientists like those at this table believe in the wisdom of the collective, the insight of research and petri dish amplification. Understood. But the cell already operates as a collective from years of evolution. We are the research collective; you must operate as an individual."

"Slow down, lil' Willy," Dr. Porter sarcastically announced to the table. "Gave the booger the floor, and he got excited."

"My name is Winston."

"Willy, Winston, whatever."

Dr. Porter was becoming increasingly uncomfortable. Globally respected for his scientific views, he had no interest hearing scientific contradictions from an imaginary cell. This was supposed to be fun and games. No real ideas were to emerge. He tried to color the atmosphere with jest, but the phantom conversation had become real. Things had taken a turn. Winston, who had now intellectually possessed its host, wasn't amused. Angrily, he had a few questions for Dr. Porter.

"Where were you?" Winston asked.

"Excuse me?" Dr. Porter sat up and responded with an arrogant squint.

"Where were you when the foundation of the earth was laid? When oceans were holes of dust and the sun a flickering amber with ambitions of greatness?"

"I'm not sure what—"

Winston asked again. This time pronouncing every syllable of every word.

"WHERE WERE YOU when the light was separated from the darkness and infinitesimal molecules were visible to the unclouded eyes of early man? I'll tell you where. You were light years from existence. You should listen. You should not speak."

Dr. Porter was in shock. He was speechless, a phenomenon no one at the table had ever heard of, much less witnessed. Winston, now calmed, continued in a more peaceful tone.

"Your body is an occupation. You don't have cells; cells have you," he said. "You are literally a pass through. Now, what passes through you is perfect in every way, which in theory means you can also be perfect in every way. So where does contamination happen? And through whom? Cells are only contaminated by what you do, what you eat, what you believe, and to what you expose yourself. Cells are handcuffed by fatigue. You need sleep because we need sleep. While sleeping I have the peace to reorder and archive the brain's memories without the pollution of outside stimuli. While sleeping I lighten your allostatic load, balance your blood pressure, prune, and clear your synapses. Sleep is nature's maintenance shop. It's where your body goes at night to be repaired. Cells are strong. In tensile, compression, and shear terms, we can be a million times stronger than graphene, and at the same time one million times weaker than one wet sheet of notebook paper. It is you and you alone to whom nature has given the power to decide what your cells will be for you. Our genome, or, to use your terminology, our intellect, is barely leveraged. The cell brain has in its reserves millions of instruc-tions and organized contingencies for every disease that many you have discovered and even more that are yet to surface in your world of medicine. As a cell, all of our information originates from natural and learned sources and how we manage trillions of operations on a daily basis comes from an extensive network, a decisions tree if you will. Each of us has around fifty trillion cells (50,000 billion). If we took the DNA from all of those cells and laid it out in a linear fashion, it could wrap around the earth two and a half million times or reach to the Sun and back 300 times.[7] In essence, you could learn a lot from us. Unfortunately, our information is largely useless because, again, seldom is the cell brain leveraged. We are researched, prodded, poked, and sliced but rarely leveraged. Why? Because as aggressive as we are as cells, we are passive in one way. You, the host, must call upon us to perform at the levels we are able to perform, which are higher than any human can think or imagine. In fact, thinking is one of the central problems. Human thinking is generally based on limited exposure to ideas presented to you in a particular sequence. You learn math, language, and other elevated forms of academics in the order of their simplest form to their most complex. Cells learn disciplines starting from their most complex form then ultimately arrive at their simplest. This is why, to the naked eye, the body operates seamlessly while billions of operations are taking place. We've learned to operate under the cover of simplicity.

Humans tend to think about things that have already been thought. On average, thinking is intellectual recycling—which rarely results in new ideas. At best, you arrive at better ways to express old thoughts. This is why belief, which elevates thinking into hope, is so powerful. Belief is an electronic

---

7 https://www.nature.com/scitable/topicpage/dna-packaging-nucleosomes-and-chromatin-310
DNA Packaging: Nucleosomes and Chromatin
By: Anthony T. Annunziato, Ph.D. (Biology Department, Boston College) © 2008 Nature Education
Citation: Annunziato, A. (2008) DNA Packaging: Nucleosomes and Chromatin. Nature Education 1(1):26

pulse the cell recognizes as nearest to our given abilities. Belief, then, is the key to superhuman accomplishment and hope is the nutrient of this belief. This interchangeable duo activates eternity within you. Only when belief is required to make sense is the pulse short-circuited.

"So you're saying belief makes everything happen just like that—*voilà*?" Dr. Porter asked his colleague before catching himself. Wait! I'm acting like Winston is real."

"I am," Winston responded. "That is if you believe I am," he continued.

"Cells in most human beings languish as genius without purpose. The human cell suffers an existence akin to rich eternity forced to wear the impoverished robe of time. We are so much more. Do you understand that for every disease that has ever killed a human being, there are millions of others who have survived the same disease? Cells are a shared collective, a biological internet if you will. Our genome holds the successful, curative knowledge saved from the survivors. You've already noticed the shared enzymes between species, but this only hints at the power of our sharing. You've already noticed that the fingerprints of a koala bear are indistinguishable from those of humans. We all share and there are millions of other synthesized examples you've yet to discover. The cell knows how to heal you.

"The cell knows how to cross-collateralize—applying ancient

answers to your body's health from pathways you have yet to study because you don't know they exist. As I converse with other cells, one thing surfaces often: the fact that man has succeeded in connecting your communication through technology should finally bring to light how cells have operated forever. All information is available to all cells. A person sick in Dallas has cells able, through belief, to access cells of another healthy person in California, and with this information heal their host. Oh, did you believe that only in your world could physicians log into a system and access patient records from across the world in real time or conduct medical procedures in absentia? Certainly, you didn't think this was a new idea. The cells that control human intellect are the ones that planted in you the idea for the internet. And what about the time issue I started with? Your scientists have provided estimates that, under ideal conditions, DNA has a half-life of approximately 521 years.[8] This implies our usefulness or trackable biology ends in about one thousand years. It is much longer than that, but one million years sufficiently makes my point. We have the ability to support the host, the human body, for this amount of time.

"In you is eternal life. Scientists would tell you our life outside of the host is based on a laboratory existence and that cells must

---

8 http://www.the-scientist.com/?articles.view/articleNo/32799/title/Half-Life-of-DNA-Revealed/

receive nutrients from a third-party source to survive, that our capacitance is too limited for autonomy. This is false. We are self-fueling, self-refueling, and self-sustaining. We have the capacity to internally create what we need although our genome may not organically house the components we create. Imagine stepping out into the rain and growing an umbrella at the end of your arm although there is no umbrella inside of you. This explains the cell. We have creative power within us beyond what is in us. This is why you, through the immeasurable, unmatched power of belief, can grow a healthier you from inside of you and have this healthier you take over for the sick you.

"Although limited by his understanding, I can applaud the innovative approach Dr. Porter used a few minutes ago, oh, and yes, cells have ears. We hear everything you say and we understand everything you think, even when you don't have the words to adequately express yourself. Porter used an ancient reference to explain the power and priority of sleep. What he referenced is more in line with transubstantiation than counter fact. Concerning the self-powering nature of the cell, I will walk the same narrative trail as he.

"In the same document referenced by Porter, there is a story of a man named Lazarus. Lazarus became very sick and died. His sister was very upset with Jesus who was his friend. Her words of critique to Jesus were, 'If you would have been here, my brother

Lazarus would not have died,' to which Jesus responded, "I am the resurrection, and the life: he that *believeth* in me, though he were dead, yet shall he live: And whosoever liveth and believeth in me shall never die. Believest thou this?" Shortly thereafter, Jesus raised Lazarus from the dead. I offer a similar refrain.

"We are your source of life without a third-party hand. Every piece, every thought, every grain of knowledge you have is because the knowledge first existed in the cell. You have life because the cell lives." The table was pin-drop silent. Dr. Porter was nearly fuming from the last couple of interactions with this invisible Winston. Porter shouted out, "But none of this is true!" Winston sharpened his aim toward Porter again.

"What is truth, sir? Better yet, what is the future of truth?" the cell asked in rhetorical tone.

"While I appreciate your allegory on sleep, there is a problem. You utilized a text with which you disagree as cover, a cover to protect your philosophical world view from critique. What you didn't know was you made an even more profound point. That is cells are creators of truth. We don't react to truth; rather, truth is the result of what we believe. The story is true to all those who believe it, maybe not for you but those who believe the story benefit from its wisdom.

"Cells were around when the Wright brothers, doubted by an entire world, still believed in something without evidence or science to justify their idea of flight. I was around to watch Martin

Luther King, Jr. fight for what was morally right although no one had succeeded before him. I was in that sobering chamber. I heard the whispers of persecuted Jews who gathered in comforting circles as billowing gas entered through metal grates. They held on to hope. They believed their children would one day live although they knew all those in the room having the conversation would die. Belief opens our capacity. If you believe, you gather the entire force of billions of us deep inside of you, a force that doesn't check statistics, facts, or even science. We existed before all of these. Belief allows us to do our best because, while we are limitless in our capacity to heal, we are helpless to make you believe.

"You may never again be privy to hearing about your health from the mechanism in charge of your health, but just in case a human cell never speaks to you again, think on these things as I leave. The world is moving so fast, everyone is in such a hurry, yet you are in a hurry to do the wrong things and go to the wrong places. Be in a hurry to become healthier. Make it a priority to sleep and sleep often. Second to belief, sleep is the most productive action you can take. Sleep early and often. Cells live as long as we do because we sleep an enormous amount. Follow our example. We've been here awhile. See you in your dreams."

☙

Ann-Louise Johnson, IFMCP, RN

The Impossible Dream

To dream the impossible dream
To fight the unbeatable foe
To bear with unbearable sorrow
And to run where
The brave dare not go
To right the unrightable wrong
And to love pure and chaste from afar
To try when your arms are too weary
To reach the unreachable star
This is my quest
To follow that star
No matter how hopeless
No matter how far
To fight for the right
Without question or pause
To be willing to march,
March into hell
For that heavenly cause
And I know
If I'll only be true
To this glorious quest
That my heart
Will lie peaceful and calm
When I'm laid to my rest
And the world will be
Better for this
That one man, scorned
And covered with scars,
Still strove with his last
Ounce of courage
To reach the unreachable,
The unreachable,
The unreachable star
And I'll always dream
The impossible dream
Yes, and I'll reach
The unreachable star

Joe Darion / Mitchell Leigh | The Impossible Dream lyrics © Helena Music Company

CHAPTER SEVEN
Motion / Emotion

A thick leather binder balanced on the edge of the dinner table. It was time to pay. It was past time to go. Dinner was life changing. The meal was over, but Winston was not finished.

There was one more item he needed to convey before returning back into what could be another half-life of silence. Without warning or transition, he drafted a third party into his monologue, a woman named Hope.

"The body is designed in layers," Winston said.

"A hierarchical order from highest to the least capacity. Capacity is relative, however. Organisms with the least capacity are still light years ahead of man's understanding. That said, there is one within us that holds not only capacity but immense execution without error.

"Simply put, she has all the power."

"Who?" Dr. Newman asked.

"Hope," Winston answered. "She is the indistinguishable fire of the soul. She is the reason the body, after losing all medical signs of life, reaches for that last breath. She is unrelenting—eternally expanding beyond human thought. Hope will not be satisfied until there is nothing left to hope for, not because failure has finally fatigued the hopeful spirit of man, but because every human being has all the health, happiness, joy, and adventure they could ever want. Hope wants it all."

Dr. Newman was ready to go fifteen minutes ago.

It seemed now he couldn't move—again.

Winston went further.

"The intricacy of Hope and how she breathes life into me, the cell, is beyond human comprehension. Science sees dimly the true glow of her genius. Her reach is beyond distance. Her ideas are beyond imagination. Most view Hope, or mitochondria as you like to call her, as functional in operation, as a miniature powerhouse generating adenosine triphosphate for cellular energy. While this is vital to your local understanding of eternal biology, this is the least of what she does. She is worlds more.

"Hope is the self-powering, self-correcting, spontaneous reproducing center of life itself. She is fuel that never needs refueling. Not only does Hope provide power to the cell, but she provides sentiment, intellect, motion, and emotion to the body. She is the axis around which the cell and its billions of independent operations depend. Hope informs the all-knowing genome. Digest this. Hope informs the all-knowing. Using the terms of your physical world, imagine this. Imagine the generator in your garage was not an emergency product but an irreplaceable source of daily life. From it, you would gather far more than day-to-day electricity. This generator would not only power your house and the mechanics therein but your mind and body. From it you would not only gain the energy to think but the capacity to think creatively. This generator would enable you and everything around you to grow. In other words, in the sky of the cell, Hope is our sun," Winston continued.

"Regenerative qualities of the cell are well studied, what isn't fully exhausted is how every component in the cell revolves around the power of Hope, the mitochondria. Do you really understand what I just said?" he asked rhetorically. "What I'm saying is this; you understand how genetic predisposition springs from the cell, but your cell has one source from which it springs. There is one source in higher order than itself. Your human concept of hope as an optimistic thinking position has its roots in the mitochondria of the same name. If you seek not this understanding, all of your research, past and present, is festivities of insignificance.

"Language itself and the complicated connotations found therein were delivered to you by Hope as a mirror of what lives inside of you. Even your infant baby carriages look like they do because the design idea was given to humans by Hope. Look again at a microscopic picture of human mitochondria. It looks like a baby carriage, doesn't it? And what one word best describes a new born baby? Hope.

"Did you know Hope is a musician, an architect, a warrior, a soldier, and a diplomat? She possesses all requisite skills to perform masterfully in any discipline. Show me a person in your society strong in any area of higher learning, and I'll show you a person with heavy doses of Hope inside of them. Competence, compassion, creativity, and life itself have their foundation in Hope.

"Anytime you experience an unforgettable moment of joy or peace that circumvents understanding, you can be sure it was Hope guiding the experience. Hope takes you beyond you. She is the coal from which motion and emotion gain their amber, an amber that becomes fire, a fire that moves upon a crowd in layers warming each one it touches. Dr. Newman, you know this well, don't you?" Winston asked.

The doctor nodded in affirmation.

What the rest of the table didn't know was that Dr. Newman and his pregnant wife had attended a memorial service for a dear childhood friend just a week earlier. Winston knew Newman had been thinking about this event the entire night and took the opportunity to explain Hope in greater detail by recalling the events of that day.

"Hope and I both were there. We saw everything in real time," Winston explained.

"We knew the day would not soon be forgotten when Hope encouraged Ms. Lula Williams, the eighty-year-old black woman, to stand to her feet. A steadfast member of the church for three quarters of a century, hers would be the last voice heard. Frail, she shuffled to the podium and gripped its corners. And with a voice soaked with tears, she sang the last refrain of a time-tested hymn.

In seasons of distress and grief, my soul does often find relief, and oft escapes the tempters snare, by thy return, sweet hour of prayer.

Silence filled the church.

"Ms. Lula then stepped away from the program and told a short story of the many Sunday mornings she would grab the deceased from his mother's arms and rock him to sleep. She did this so the congregation could hear the preacher and not just a crying baby. These words relaxed the congregation and provided a much needed laugh."

' "But I had a little secret,' Ms. Lula told the congregation in a whisper. 'I would sing a little song he loved.' "

"Ms. Lula looked over at the organist and motioned for him to sit this one out. Then she sang just a few words from the song that always brought peace."

Dr. Newman excused himself from the table.

He had witnessed all of this firsthand. Hearing everything repeated back verbatim was too much. Firsthand he saw motion, emotion, and the power of Hope move through the audience like a tidal wave. Layer by layer, row by row, all were affected by the shaky authenticity of a woman who had seen her share of sorrows and therefore could, with surety, encourage all in attendance that they would survive this emotional day.

Hundreds traveled from a world away to be in that church. They all came to celebrate the life of a father, a soldier, a confidant, a friend. Local media picked up the story. A few cameras littered the back corners of this rural sanctuary. Only a handful of times had this little town had one of its own, a decorated veteran, fall on foreign land. The service was nearing its end. Television crews began rolling up cables, but then something happened. This is the real reason Dr. Newman had to excuse himself. He didn't want to hear Winston recall what happened next.

"You see, this soldier's son, in a guttural act of pure emotion, slid out of his front row seat, stumbled to the face of his father's casket, and fell to his knees. His arms, weakened by grief, draped its cover. The room, already still, now stopped breathing. Camera crews dropped their cables. Ministers hid their faces in open hands. The little boy was oblivious as to how his motion connected everyone. This was his father. They would never throw another baseball or take that weekly trip to the local ice cream shop. Never would they have the opportunity to make memories for life, and so, today, this young boy would…make one last one. The picture traveled around the world faster than the internet could take it. A still shot of a son, in a blue suit he had outgrown, on his knees gripping the casket of his father, his hero. Even today, without knowing the story, without knowing the family, or any details, the picture reaches into you as a conduit for the human network of emotions. We are connected. We are daisy-chained, networked together. We are designed in layers.

"We connect through the transferability of care. Like melted steel, humans are firmly welded together. Although much is made of our differences and disagreements, connected we remain. Just like the human body, we come together in layers from country to community, neighborhood to individual house. And while these organic, default connections exist, how much we benefit from our natural connections depends largely on our belief that the connections exist. Once we believe, it is then our intentional activities to make scalable what is natural that can change a community and everyone in it.

"Think of that picture again of the little boy. While most human beings will feel the emotions emanating from it, there are gradations or differing levels of how deeply one feels the photo. It can be safely said that a person who has experienced the same pain as the boy through the loss of a loved one would feel, or understand, this picture better than one who has not. It can be safely said that an individual who has been forced into the dark night of burying a loved one who left for war vertically but returned from war horizontally may empathize at a deeper level. In other words, these persons carry a double portion of the "gene" called understanding.

"A person's genotype represents that which is embedded in their very being, while a person's phenotype represents their genotype plus those things observable to the naked eye such as height, eye color, and length of hair. Phenotype is influenced by environment, while genotype is influenced by progeny. There is a fascinating discussion in the July 26, 2007 publication of *The New England Journal of Medicine*. In it, Albert-László Barabási, PhD summarizes the relationship between what is genetic and what is influenced by a person's surroundings. The article takes a look at obesity; in that while the probability of obesity is genetically born through the *FTO* gene, and that this probability doubles depending on the carrier having two *FTO* genes, there is another association that is just as powerful, if not more—our social network.

"It was found that if a person enters into a friendship where one person becomes obese, the other friend's probability of also becoming obese is increased by 171 percent. This is a greater propensity than the genetic bias of being born to obese parents, even if the obese person is only a friend of your friend, your chances of blossoming beyond what is healthy increase by forty percent. This transference of unhealthy activity is eye-opening because a person can become friends with someone who is morally sound, intellectually sharpening, yet their physical influence may not be best. In other words, a person's influence can be a combination of good and bad, which encourages us to look with a sharper perspective at who our friends are and, furthermore, who their friends are. But a word of caution: weight is in no way a direct determinate as to the quality of an individual. It is obvious, but worthy of saying. Our true friends, family, and those we respect come in all shapes and sizes. Integrity and honesty don't show up on a scale.

"Networking, a common term today normalized largely by technology, means more than passwords for shared access to documents. In genetic terms, networking allows us to benefit and be pained by everything and everybody connected. A virus and an antivirus spread via network. Which are you receiving? That is what level of health or sickness is being transferred to you because of who you are connected to? To this point, *The New England Journal of Medicine* makes another interesting observation known to those who study the power in the cell,

" '...network analysis is poised to play the biggest role at the cellular level, since most cellular components are connected to each other through intricate regulatory, metabolic, and protein-protein interactions. Because of these many functional links, the defects of various genes spread throughout the intracellular network, affecting the activity of genes that otherwise carry no defects.' "[9]

༓

Dr. Newman never returned to the table. He met his colleagues back at the hotel. The night had travailed from offense to intrigue to inspiration to emotion—so many emotions. All of this was too much. Dr. Newman called his wife, laid back, and reflected on the day. He was left with profound scientific, political, and practical questions. The idea of hope being a biological force stuck. In a departure from what was usually his purview, he began to think about hope in a different way. He started to think about hope from the inside out, from the bottom up. He wondered how we as a society turn hopeless communities around. How do we turn what is biological, what is personal, into what is scalable? How do we take what we feel inside and make it something everyone outside can observe and benefit from? How do we encourage others to tap into what is already inside

---

[9] Barabási A-L, Oltvai ZN. Network biology: understanding the cell's functional organization. Nat Rev Genet 2004;5:101-15.

of them from birth? The way Winston spoke, albeit harsh, was maverick. Hope, for Dr. Newman, had just transformed into a superhero. The only difference is this superhero doesn't emerge from dark corners to save the public. This superhero is inside of each person.

This isn't lost on everyone. There are communities around the country that promote hope as a lifestyle. These communities have materialized mitochondria in physical form, and thereby power our imagination, energy, and physical health. Communities like these are designed in layers.

☙

Visit Vancouver, Canada or Seattle, Washington and you'll notice something. As soon as you exit your car, you can feel positive community spirit—a tangible cooperation among the residents. The benefits from high-level planning filter down. Stand on a random street corner in Seattle and you'll see people playing, walking dogs, biking, and unloading canoes for an afternoon excursion. This not only encourages people to move but to move vigorously, a society of mitochondrial motion.

This is important because motion begets motion. Sitting is the new smoking. There isn't much that harms our health more. In the absence of motion, the body can enter into a kind of living rigor mortis. Motion gets rid of inflammatory molecules that can lead to disease or pathologies. Exercise is life. The mitochondria, hope, does a push-up each time you do a push-up. Your physical body and its activity are mirrored down to the cellular level just as the cellular level is mirrored upward into our genetic and phenotypic display. When you smoke, your cells smoke, when you sit still, your cells and mitochondria sit still, but when you exercise so do they.

Our emotional makeup, our emotional intelligence, is as simple as it is complex. Motion and emotion feed off of each other in a way that can only be described as conjoined twins. Conjoined twins exist because of each other. Independence would kill them both. Heartbeats are shared, sickness is suffered together, and health collectively enjoyed. The circulatory system of two babies becomes one. If one dies, so does the other. It is this shared network of life that makes conjoined twins so intriguing. As an interesting side note, genetically identical twins are most commonly female, just like hope. This is because female conjoined twins survive outside the womb at a rate three times higher than their male counterparts.

Simply put, your emotions and motion are one. Have you ever noticed how difficult it is to stay depressed while exercising? Have you noticed how perspiration often encourages inspiration? Everything is connected; even if the emotions and motion aren't your own, they still move you. Observing another person emotionally engaged has a tangible effect on all those who watch. This is because spirit calls to spirit, and deep calls to deep. We are connected.

☙

There is something else that affects our emotions in a not so obvious way: our food. Did you know there is happy food and sad food? As Dr. Candace Pert points out in her book *Molecules of Emotion*, most molecules of emotion are made in the gut. If our emotions are derivatives from the gut, what we put in our gut becomes extremely important. At the risk of echoing elementary ideals, garbage in, garbage out.

So, are you eating happy or sad foods? Sad food can be found in the central aisles of your grocery store. These foods are high in carbohydrates; they are highly processed foods that pollute our genes in so many ways. So when you shop for food, remember you are also shopping for emotions.

In what seems contradictory at first glance, our best motion can be stillness. Our best motion can be no motion at all. However, stillness and inactivity are very different. As we discussed earlier, sitting (inactivity) is very unhealthy. However, purposeful stillness or intentional physical peace can do wonders for the body, mind, and spirit. Waiting with purpose, which essentially allows us not only to show strength outwardly but to gather the forces of strength within us, is powerful. I've always appreciated the explanation of what waiting accomplishes from an Old

Testament perspective: *But they that wait upon the Lord shall renew their strength; they shall mount up with wings as eagles; they shall run, and not be weary; and they shall walk, and not faint.*[10] I've learned it takes power and courage to be still. While speaking around the country, I've consistently heard one observation, which is the most powerful portions of my speech when I'm on stage are when I'm perfectly still. In a world of perpetual activity, stillness causes attention. So good emotions are not just represented by smiling; they can be quiet, serene, or even melancholic and still be healthy.

Three months after a life-changing dinner in Atlanta, Dr. Newman and his wife welcomed a precious baby boy into the world. As the nurse handed him the warmly wrapped child, she asked a question.

"What is his name?"

Dr. Newman, full of emotion, answered, "His name is Winston."

---

10 Isaiah 40:31

CHAPTER EIGHT
Support

Hope is a master surgeon able to operate on two areas the field of science has yet to conquer: the mind and the cell. It is the mind, its operating premise, and its belief system that most profoundly affect the human body. What we believe, what we perceive to be true has the power to change what is actually so. And belief, when fully codified, is surrounded by a community of support, a community of individuals who also believe, a jury of sorts who are committed to convicting you to living a full life.

Human support systems, belief teams if you will, inspire us. To perspire is to sweat on the outside, but to inspire is to sweat on the inside. And once this internal inspiration partners with outside support, we run longer, reach higher, and dig deeper because we are no longer limited by personal capacity. We are powered by a collective, powered by people who invest their energy into increasing yours. This is why the marathon runner, whose physical body is depleted and whose mind is fogged by exhaustion, can be gifted another mile through the innocent voice of a young child yelling, "Don't give up!" The collective is strong when we are weak, awake when we are asleep, and energized when we are fatigued. The collective allows us to safely take risks.

Those three words, "Don't give up," can make us believe where once we may have doubted. Our belief system is the container into which we pour all information. When we see odds against our recovery or statistics that nearly predict our failure, belief can be the last line of defense. What we believe colors everything we hear and see. Our "belief container" must be solid enough to hold to convictions yet porous enough to consider better ideas, which brings me to my next point. In my years of practice working with the vibrant and healthy and the ill and the terminally ill, there have been times of inexplicable recovery. I've witnessed recovery events that directly contradict the integrity of all we know scientifically. As scientists, our guttural reaction to something that moves from the intangible to the tangible is to pass the experience off to a random, natural anomaly much like a hermaphrodite—a hiccup of nature that even if unexplained doesn't disturb the order of things. What you believe determines what you have the capacity to accept and even use to correct old assumptions. Correcting what you previously believed to be correct can be risky business.

It can be risky to grow new horizons anywhere but especially if the horizon is inside of you. By default, new horizons must compete with old, established parameters of thinking, and zones of comfort tend to zealously keep watch over anyone attempting to blaze new paths. But the one who finds courage to challenge the status quo finds something rather interesting. Many people will suddenly appear and be the underpinning of your new initiative—the initiative to change what needs changing. These are individuals who dare not be first in such a bold move but are willing to be second in supporting your boldness. However, if you make timidity your home, you decay, and you decay alone.

It takes courage to stand up when millions around you are seated. It takes fire to tell your family and friends you are going to survive a sickness no one before you has. You will disturb relationships when you decide to change your living, eating, and thinking habits because others have made peace with the sweet fruit of mediocrity. You see, people generally don't believe in your dreams until they at least have experienced a belief in theirs. This has direct relevance to your health. When a person is sick, proper support requires more than just qualified health professionals. Proper support requires that everyone in your sphere of influence is of a certain mindset, the mindset of visionaries and dream chasers, optimistic personalities who refuse to go away quietly in any area of their life. You need people who refuse to be refused.

You need a team that not only accepts the pragmatics of your sickness but holds greater to the surety of your full recovery. Support and hope can take on the oddest of forms and be found in the most unlikely of places—even at the stoplight.

☙

I learned her name was Debbie. She was just standing there—still, stoic, sad with a hint of optimism. The unseasonable heat and humidity were made worse when mixed with exhaust fumes from hundreds of cars idling in place. I was in one of those cars. The white truck. True to the life of a busy professional, I was in a rush to go somewhere meaningful. I didn't have time to stop, but the red light stopped me, and out of my left peripheral view I saw one word that seemed out of place for a standard beggar on a standard street corner. Her cardboard sign had one unique request. She asked for hope.

My curiosity called her over. She didn't smell. She was just hopeless. Her smile was a hybrid; the memory of better times could be seen cutting through the despair of the present. She didn't overtly ask for money. Debbie wasn't greedy; her sign kindly asked for someone to grant her hope for today, implying that she knew not to expect anything beyond that. Debbie was, in reality, seeking the support side of hope, the portion of hope that enables us to keep our carafe of encouragement full to make it through one more twenty-four-hour cycle.

Support breeds confidence to keep red the ambers of hope. Support says someone cares enough not to let us fall. Support is patient. Support is nonjudgmental. Support can, at times, originate from people with whom we have no personal relationship. Support, in its purest, most authentic form, has no other objective but to do that—keep a person or thing from bending under the weight of outside pressure, pressure that wants you to change from what you were perfectly designed to be into what is most comfortable yet most damaging.

When we think about the uses of support through a mechanical lens, a couple of interesting facts come to light. Support beams are successful when they resist. A bridge that maintains its integrity over time does so because it has opposition preventing it from deforming away from the architect's original form. In other words, true support prevents you from going in any direction that is counterproductive to your best self. Human support is no different.

While on one hand we humans tend to migrate toward those who look like us, think like us, and enjoy the same things we enjoy, that is not always the best form of support. In fact, more often than not, these types of relationships can become enablers of our fall. People can usher us right into our demise all under the cover of socialization. What we need is positive opposition. Our health demands it. When we join a gym, we willingly place ourselves in partnership with opposition that will maintain a standard, an uprightness if you will. It is the function of unyielding weight and respiratory exhausting that ironically increases our respiratory capacity and physical strength. Positive opposition. People in your circle must be dedicated to your success even when you are not, dedicated to keeping your standards firm, the standards of mental, physical, and emotional health. Your standard is just that—what is best for you, your body, your skill set, and your individual capacity.

Debbie on that street corner on that busy day did something I never expected her to do, and this strikes at the center of hope and its transferability. She encouraged me. Debbie gave me hope. How did that happen? I was in my car enjoying the air conditioning of a modern vehicle along with leather seats and the pleasant sounds from a high-end audio system, but hope is infectious—communicable. So powerful is hope that the very request for hope can deliver hope to the person being solicited. The experience was as confusing as if a homeless man, having none himself, built a home for another person; as if a naked man unclothed for cold weather sat down and sewed a warm coat for another. The asking for hope is in and of itself the giving of the same. I placed a monetary gift in her hand, but Debbie, this young woman who had released pride from her eyes in order to stand in that unkept, grassy median, gave me more. This young woman who somehow understood that the magic of hope was far more valuable than wrinkled dollar bills from drivers seeking to free themselves from the guilt of a more blessed plight; this woman, Debbie, reminded me of hope's power. The multifold nature of hope and its support mechanism can both uphold a bridge and Debbie from falling any lower.

For all we've learned about hope, its connection to motivation and its dominance over reality, there is still another piece without which hope cannot be all it can be, and this unique role can only be filled by support. Support isn't in play until it is needed. Let's return to our mechanical example; support beams are basically pillows that transform into unyielding resistance once pressure is applied. When no pressure exists, support is essentially dormant a sleeping elephant, inactive with enormous potential. This is why many who are sick never realized their true friends until they become sick. Challenges are excavators. Challenges let us know what's under all those smiles and nice wishes. Challenges show us where the diamonds are buried. Many who promise never to leave are the first to do so, while others who seemed only peripherally interested in our lives can become lynchpins of support and friendship once the pressure of an ailment arrives at our front door. Some people are best in bad times.

଄

One wintry day in Pennsylvania, I had the unique experience of meeting a client in an unusual way. I collect quilts, and so does she. The rest of the story is accounted here in a lunch conversation I had with a friend.

The man's booming voice could be heard without microphone.

"$155! $155! $155! Is there another bid for item 3497? A warming quilt you'll wish you had. The snow angels are coming; the snow angels are coming! Made with love, made with care, made for one person—you!" the auctioneer shouted in a southern preacher's accent.

"$155, $155, $155 do I have..." Anna raised her hand.

"Boom! Bam! $200 thank you, ma'am. Do I have $225? $225 bid; $225 for a one of a kind winter quilt made for this and generations to come. Your family will love it! If I weren't ineligible I'd buy this thing myself in a heartbeat!"

One lady in the back of the room leaned forward and, with a directed squint, raised the bid to $300.

"$300 bid!"

"Do we have another? Item 3487, a beautiful, beautiful... Got it! Right there. $350 back of the room."

The $300 bidder stood on her tiptoes to see who was outbidding her.

"We have $350, $350, calling $350!"

"Is there another bidder?"

Pause.

A creeping anxiety populated the room. A mix of professional courtesy, personal kindness, and a smidge of selfishness.

"$350 going once, going twice. There it is! $400! You! $450! We got us an auction, folks. The horse is riding heavy and high, this just might go down to the wire."

The auctioneer could feel it. The tension in the room was rising fast.

"Item 3497, $450 it is. We're rolling like the Pocono Mountains! East! West! North! South! Don't know where this thing is going, but wherever it is, we're going fast! Who takes the quilt home tonight? Folks, when it gets cold, don't say I didn't try to warm you! With all due respect to my wife, of course, but I digress."

Terry burst into laughing.

"That guy cracks me up," she said.

The auctioneer continued.

"$450 went once, it's going twice. Three times a lady, who wants in?"

The two women now found each other's eyes. Both ladies were hedging to see who wanted the quilt more than the other. The auctioneer hit the wooden auction desk with a judicial snap and shouted out the final call.

"$450 going once, $450 going twice—ssssold to the nice lady in the back!" The audience of nearly fifty dedicated bidders applauded as the winner wearing a nearly apologetic smile made her way to the front to collect the old-fashioned quilt. Anna dropped her head. She really wanted that quilt. She wanted it not just for warmth but for peace of mind. The quilt reminded her of her grandmother, but with all that was going on, she needed to watch her dollars. The quilt, as simple as it may have been, was going to give her a little piece of peace. She was going to read with it, cuddle with the dog under it, and just try to bring as many memories of better times back into her life. Maybe she'd win next time. Maybe. Anna filled her cheeks with air and closed her eyes. The auctioneer moved on.

"Our next item is one 1945 work table with the tools to match!" Anna grabbed her purse and headed for the door.

"Good bidding," a voice says from behind her.

"Yes, nice auction. Congratulations, Johnson," Anna said to the woman, reading the name tag directly off on her lapel.

The woman reached into her purse and pulled out a business card.

"Here's my card. God bless you."

Anna stood in place for a moment not really knowing what to make of the brief, yet nuanced conversation.

The woman turned and walked away.

Anna placed the card in the bottom of her purse and turned her attention back to the issues facing her: diabetes and severe nerve damage to her feet. The next morning she sat in front of another specialist. She was trying to find relief. She found herself explaining her story again to another physician, but she was getting the same results. Medicine was prescribed for the pain, but Anna still had no understanding of why she was suffering or what she was suffering from. And it was during this time, as Anna was trying to get a handle on her feet and the diabetes, that she was handed another blow. She learned she had breast cancer. Anna, a nurse practitioner, understood medicine well, but even for her, life had become too much to bear. In the end, Anna recovered. She won as a result of intense support. How did she win? Many years later, during a 2008 conference on the importance of community support during cancer recovery, she was invited to be a guest speaker. She described her experience during her time of sickness this way.

After forty years of marriage, I divorced my husband—this will become relevant in a moment. Even as a nurse practitioner, thinking about surgery for breast cancer filled me with anxiety. I didn't know what to expect. It's so much easier to give directions and advice when you aren't the patient. But then I remembered something. A friend of mine had surgery for kidney cancer and it amazed me how quickly she recovered. I needed to speak to her and find out how she came through with such flying colors. We met up for coffee and she handed me a card.

" 'Her name is Terry,' my friend told me. 'Call her today. She puts together an unbelievable team to play this game called Adventurous Health, and her team is undefeated. It may sound like exaggeration, but the woman is the real deal.' "

I must have had doubt, confusion, or both on my face because my friend took her explanation of how Terry operates to the next level of persuasion.

" 'Let me explain,' she said. 'Terry thought of everything. Her team covered me. She called a meeting of over forty people from all walks of professional life, and right there at the meeting people signed up for me. They signed up to make sure I would make it. And I say *make it* not just in the sense of recovering from surgery, but they wanted me to come out of surgery stronger than I went in. What I'm trying to say is, I didn't simply return back to my original health, I returned to a state of health that was more vibrant and more resilient than before I was sick. Most people think of gaining their strength back after surgery. Terry thinks of giving you more strength than you've ever had in your life. She taught me that post surgery isn't a platform for recovery but a platform to cultivate greatness. Let me tell you about her team. Someone is responsible for making you laugh, yes, joy is a part of strength. Another person feeds you positive thinking on a daily basis without books, audio books, emails, phone calls, and the like. Someone else plans and fixes your meals, another person looks out for your appearance to make sure you not only feel good but look good, which in return makes you feel better. An individual manages and administrates the team's activities while another person handles transportation, getting you back and forth to your appointments. Whatever an individual does in their professional life, whatever their area of competence, Terry leverages this skill on your behalf. It is an ingenious plan. Not a plan to survive, but a plan to thrive.' " I looked down at the card, I looked up, then I looked down again. I was shocked at the information I was hearing and the name I was seeing. This was the same card in the bottom of my purse I got from the auction.

"This is who helped bring you back to health?" I asked. 'But... but this is the quilt lady!' I whispered.

My friend ignored my question. 'Call her,' she said. I walked outside and did precisely that. The phone rang once.

"Hello, this is Terry."

"Hi. My name is Anna and I need to make an appointment."

"Sure Anna."

Terry took my information and scheduled a day and time to meet. I got off the phone, sat back on my couch and cried. I'd never felt so alone in my life. First, my marriage fell apart, and now my body was taking a hit. My husband and I had been together for literally a lifetime. Often, I wondered if the breaking of our relationship had anything to do with the sickness I was suffering. I'd heard all the ideas about one's immune system being weakening when emotional stress becomes too much. We were such a team. Even when things were bad, I knew I could depend on him. I just sat there on the couch for about thirty minutes. The moment I went to get up, the doorbell rang.

Who could that be? I thought. Now was not the time for visitors. I opened the door and there she was, Terry, the quilt lady. I didn't know what to say.

"I didn't know you were coming."

To which Terry replied, 'I know you're hurting."

We stepped inside the house, sat down on that couch, and cried together. Little did I know, the healing had already begun. She asked me what I enjoyed doing, what made me feel alive.

"I love tennis," I said. "It's one of the only things I do that makes me relax."

Terry looked over at the racquet in the corner peeking out of its cover.

"I used to be in a women's league. Really miss those girls."

I showed her a tennis shaped card they sent me. It had a scoreboard in it with me winning in straight sets over my opponent.

"Really love those girls. When is the next tennis tournament?" Terry asked.

"Not for about eight months," I replied.

"Good!" she said. "We'll be ready then. You will play again." Within a few days, and after Terry had listened to the long version of my story, we pinpointed the areas in my life that needed to be addressed, areas we needed to attack head on without hesitation. We discussed why I needed to eat differently. To heal properly, you have to eat food that is alive, not food that's been killed. She got me to exercise and explained why simply popping supplements wasn't going to cut it. Essentially, I needed a new life, a life that promoted life. Terry pointed out that my genomic triggers for health were food, support, motion, and emotion.

If I was going to recover in a way that was sustainable, the most important trigger for me was support. I needed a support system, a belief team. And just like my dear friend who was Terry's client before me, just as she explained, Terry put together a team. She worked with my son. He agreed to be my exercise coach. She then made a list of things needed and a list of people to provide these things.

Everyone was given an assignment. My house became an assembly line, but instead of making widgets, we were making health. It was unreal. It was inspiring. It was life-giving. Even the lady downstairs in my apartment building said she would make jello. She was a retired hairdresser. She would bring jello then fix my hair. But I learned quickly how much I was not in control of the entire treatment process, and so I decided to use my hair as my only point of control. I cut my hair down to a crew cut before it fell out on its own. It was my way of saying I still could control at least something in my life.

Terry told me once how mushroom soup would help my healing process—shiitake mushrooms to be exact. She actually had a class at her house, had my three children come over, and taught them how to make this soup. My son actually brought me several gallons of it all at one time. Like a Six Sigma, Black Belt project manager, Terry even got my sister-in-law to make organic dishes for me—my ex-husband and daughter juiced organic fruits and vegetables for me on a daily basis. This was just incredible! Their perspiration on the outside inspired me on the inside. Terry left me with no excuse to fail.

In a very short period of time I'd gone to an auction, lost the auction for an item I really wanted, met the winner on the way out the door, got sick, spoke to a friend who then reconnected me with the auction winner who is now saving my life with reckless abandon! How did this happen? But it gets even more unbelievable.

Post-op, one day after surgery in the recovery room, I got up and walked around. This, I found out later, is unheard of after mastectomy. It usually takes three people to get you out of bed. The team had taught me breathing techniques as a trigger to get rid of anesthesia and it worked extraordinarily well. I'd been trained like an Olympic athlete whose sport was surgery. In essence, I had what we called a "drive-by mastectomy." I was in control of my recovery.

I remember the nurse coming in and asking me, "You're pulling yourself up in bed?" "Well yes," I said.

Terry had already prepared me.

'Here are the core muscles you need to condition,' Terry said.

It was really cool because my arm muscles were well prepared. I could just pull myself up in bed without bothering the areas that went through the surgery. About the only thing Terry couldn't prepare me for was the bandage—oh, that bandage!

The bandage was almost worse than the mastectomy. I had no trouble with the mastectomy, no trouble whatsoever, but my back was totally open, and running, an open sore, and I actually had to have in-home care from a company called Visiting Nurses for the first couple of weeks because that little cutie-pie, short-skirted nurse just ripped it off. She did not read the chart that clearly stated I was allergic to tape and to be gentle taking off my bandage, and I suffered her thoughtlessness.

Anyway, when I had the mastectomy, they checked the lymph nodes and they said all the lymph nodes they took were running over with cancer, so I needed to have an advanced chemo regime, and I had to have a Mediport put in. I still wanted everything to be as natural as possible, but that chemo made me deathly ill

Terry went with me to the first two meetings with the oncologist and then the oncologist said, "Are you ready to go? We're going to skip down the yellow brick road," and that's exactly what we did. My doctor took my hand and we sang down the yellow brick road and skipped down to the chemo room where everyone was sitting there looking so damn depressed. We sat there amongst absolute depression, but Terry, my team, and my doctor were filled with life and joy. We literally brought sunshine into the room.

When the team originally met, it was my ex-husband who volunteered to head the transportation service and get me everywhere I needed to go. After surgery, he continued. He started taking me to chemo treatments. It was very interesting to say the least. I was so nauseated, so tremendously ill at some point, and then I found out that I could get an injection that would take away all of that nausea. They don't tell you these things.

As soon as I got home from the first drip session, I realized I was going downhill. I lost feeling in my limbs. I couldn't pick up a pencil, couldn't do anything with my fingers. My feet lost all sense of feeling. One of the things Terry told me was to increase my circulation, to drink fresh juices and water, and also to move through the pain by writing the alphabet in the carpet with my feet. Again, it was another time when Terry just came by with things nobody tells you in the doctor's office or at the chemo room.

My experience was just a really unique one, almost an exciting experience, if you can imagine having cancer and having surgery as an exciting experience. It was so cool because I had the support and then I discovered when I had the surgery that it worked, that I actually had energy post-surgery. I was back to work a few days after I got home from the hospital. We were doing some sort of a business event, and I showed up. It was absolutely amazing. This is what I've learned. When you are sick, you quickly realize the parameters and the limits of medicine, science, and human intellect. These things can only take you so far. Once you get to that dark place, if there's not somebody in there waiting on you, you'll have some issues. It is one thing to talk about research and science—and all of that is wonderful and great—but the spiritual becomes more tangible the closer you get to surgery. I've heard people say get to know God before you need Him, but the wonderful thing about God is even if you come to Him at the last moment, He credits time to you as if you had been in with Him all along.

I've heard people say nothing good comes out of situations like this, but my ex-husband and I were remarried on May 5th, 2013. It would have been our forty-fourth wedding anniversary. I actually like him now. Miracles really happen, they really happen! I can tell you exactly when ours happened. We were headed to our third chemo session. I was feeling especially bad that day. After vomiting all morning and having extreme aches and pains in my joints, I finally got dressed, moved to the couch and waited to be picked up. It's difficult to explain, but sometimes you feel everything, not just what's going on in your mind and body, but what's going on with those around you. My ex-husband's name is Chris, and when he picked me up I could feel he was carrying something heavy. Our conversation on the way to the hospital that morning went something like this. Chris started.

"Can I ask you something?"

"Yes."

"What happened?"

"What do you mean?"

"What happened to us?"

I could literally feel my stomach beginning to churn. Chris doesn't like to talk. That was one of our biggest issues during the marriage. I would talk, he would say nothing, and things would get worse and worse. On this day, he was talking. He continued.

"I've just had a lot of time to reflect on who we are together and who we are apart. We are better together, but I don't know how you feel or where you are at this point. I've said and done a lot of things I'm sorry for. I haven't been the man I know I can be. Honestly, I really don't know how I've made it this long without you. Did you ever think of me?"

By this time I had turned my head toward the window so he couldn't see the tear stains gathering on my shirt. I kept looking out the window and answered his question in a whisper of a voice, which was all I could muster without totally breaking down.

"I...I thought of you, every second of every day."

Chris was silent. I could feel he was surprised. Our breakup was so bad he probably thought I was happy to be gone.

"You thought about me?" He asked in a quiet tone.

"I'll never stop. Time away has showed me there are only a few people you can depend on in your entire life. You are at the top of my list."

I began to cry uncontrollably as I couldn't hold it any longer. Chris never knew how scared I was at the doctor's office the day I found out what I had to go through. I just wanted him there with me.

"Chris, I knew when you found out I was sick you would come for me, and you did. You came for me."

"And I don't want to ever leave again," Chris said.

I could feel the car slowing down and I knew we weren't at our destination, but I kept looking out the window. Chris pulled onto a gravel road and put the car in park. Slowly, I turned my head to find his face as full of tears as mine was. He placed his hand on top of mine, and with my bald head, and with sickness thick in my eyes, he kissed my lips passionately.

"I love you."

"I love you more," I said.

We embraced and cried for at least ten minutes. I felt us healing, literally. I felt our bodies, minds, and spirits healing in real time. Chris pulled away, straightened his jacket and opened the car door. I didn't know what he was doing. He walked around to my side and opened my door. His hand was clinched. He looked down at it and so did I. He then got down on his knee and opened his palm revealing a diamond ring.

"Will you marry me again?"

"Yes, yes and yes!"

ಔ

Months after surgery, after I was back in the rhythm of life, Terry arranged a dinner to celebrate my recovery. It's so encouraging to gather with people who love you, who celebrate you, who want you to win. My support team deserves all the credit. But of all the things that reached deeply into my very soul during this entire process nothing compares to what I saw when my husband and I walked in that dining room with all my friends seated. Right there, covering the most beautiful table decoration you've ever seen was the quilt. Yes, that quilt. Terry brought it to my party to celebrate me. She looked at me and said, "This is yours now." So the thing I wanted to help me feel better ended up being the thing that helped me feel best. I received what I needed most—

love and support. That is hope in its purest form.

Oh, and, before I forget, Chris and I follow a prescription I believe all should follow who want a love-filled relationship: three hugs a day, because hugs can heal before you're sick.

Who is your support team? Do you have a group of family and friends who refuse to let you be less than what they know you can be? Think about Debbie, think about that bridge, think about the limitless power of love and commit from this day forward to stop operating alone. We all need help.

CHAPTER NINE
Breath

What if through breathing someone else's air we could not only acquire communicable diseases but communicable health? Most would agree that exposure to better thinking increases the prudence of our own. The opposite is also true. We discussed in previous chapters the elevated propensity to become obese when an obese person is within our network of friends. In other words, secondary smoke isn't just relegated to smoking. Secondary smoke is an idea, a principle, a law, but so is secondary health. Good and bad health are transferable.

So just imagine for a moment that cardiopulmonary resuscitation (CPR) could be used for hundreds of people at once just as effectively as it is used from one person to another. What if, like hope, breath was a cell constantly reproducing itself, constantly enlarging its territory, picking up strength, energy, and emotion from the exhalations of each person it passed? But we know this concept already; we just have to make the connection. A storm cell is an air mass that contains up and down drafts in convective loops and moves and reacts as a single entity, functioning as the smallest unit of a storm-producing system.[11] Once the storm cell is large enough, it can affect millions of people wherever it goes. Can we apply the "cell" language of storms, and the compounding power found therein to the nonvolatile air we breathe every day all around us? If so, why haven't we, and if not, why can't we? One answer is perspective.

You see, it takes courage to correlate previously uncorrelated factors. This is the most pronounced challenge of medical science today. Most pronounced because such concessions rub against the guardians of specialties—those with vested interests in maintaining their medical sequester. But what we eat, how we think, how we move, and how we breathe inextricably touch. Holistic living isn't a specialty of the nutritionist; holistic living is a specialty of those who desire to remain alive.

The genomic triggers discussed in this book represent a health collective, a wellness seesaw of sorts unable to disconnect its destiny from the weight on the other end. While everything is connected, not everything is connected equally. Some actions cannot be ignored without unredeemable consequence. One who works out several days per week can, without much

---

[11] Wikipedia

deterioration of fitness, take a couple days off, but breathing is and will always be a function of now. In breath, there is no later. As important as sleep is, it remains second generational as one can last only seconds without breathing. But does breathing simply represent the method by which we oxygenate and stay alive? Some would argue that oxygenation alone is enough, but aren't there other functions just as important as oxygenation? Have we made the proper connections?

Breath can be divided into three parts: how we breathe, what we breathe, and for whom we breathe. When we don't breathe deeply enough, this shallow breath affects our mood, limits our digestion and can become a liability to so many areas of our everyday life. Watch a person hold their breath for thirty seconds and notice how their eyes begin to bulge, facial muscles become stressed, and their veins articulate. Now, if you extend improper breathing over smaller units throughout an entire day, one will experience the same effects—just in slower increments. You have seen individuals who walk around breathing but look as if they aren't getting enough oxygen or at least not enough fresh oxygen. Oxygen and fresh oxygen are very different. Often, people are more careful with what they put in their cars than in their bodies. Second only to not breathing at all is the danger of breathing toxic, stale, polluted air. Remember, our bodies will give us what we give to them.

Breath and motivation have an intimate relationship. The critically ill have been known to live for days, weeks, and even months longer than expected because their love for a loved one kept them around. Having a reason to breathe seems odd because no one thinks about such things day in day out. Keep in mind that it is our mind that controls our body. Living a life of meaning for those we love affects the stability of our breathing as our body conforms to something we consider greater than ourselves.

There is another point that must be mentioned, which brings us back to the importance of perspective. Nothing that has benefitted humanity in dependable repetition has done so without us first identifying this thing as a potential benefit. Our expectancy, our *perspective*, sets the parameters within which we live or the parameterless horizons we have the courage to explore. We invest in that which we view as having the potential for great return. No other thing on earth is so vastly valuable as the air we breathe. Each time we take a breath we are digesting limitless potential for good and bad. Anyone in a hospital walking through an infection control area understands this well. When we wear masks over our mouths in the hospital, what are we really saying? We are saying air that passes through one individual has the ability to pick up what is in that individual, good or bad, and enter into us. We understand blood infusion, but what about breath infusion? Could it be that breath has more healing qualities than we currently understand? Could it be that the usage of masks to prevent exposure to certain diseases also has an opposite benefit of proactively exposing others to health? There's no better place to discuss breathing and its importance than with those who breathe for a living. Now, you may ask,

"Doesn't everyone breathe for a living?"

In essence, yes, but there is breathing and then there is *breathing*.

My friend has suffered from chronic asthma for over forty years. His respiratory condition places his existence under constant threat. There aren't many diseases that have as their core symptom the theft of air. From such robbery one cannot recover. So sensitive is my friend toward air and its availability that he can spot another asthmatic a hundred feet away, even if the fellow sufferer is perfectly healthy and not having an attack. I asked how he could pick them out of a crowd. This is what he told me.

"Those who can't breathe take each breath with a greater thankfulness in their eyes. If you look real close, you can discern increased suction in the Adam's apple, and there are fine, almost undetectable vascular grape vines in their eyes. Lastly, the vertical lift of the shoulders during inhalation gives their asthmatic membership away."

"Who pays attention to stuff like that?" I asked.

"Someone who values each breath."

The DNA of Hope

It is odd, but some people don't value the air they breathe. It is the ultimate irony. The very thing that no one can live without is the very thing most people overlook. More than that, breathing as a necessity of living is only a small portion of its story. Breathing is so much more. Breathing is a noun, a verb, a condition, and a skill. Breathing represents success, failures, and emotions. But most of all, breathing represents hope.

Miami

My heart jumped into my throat. It felt like I was trying to swallow cotton. How could this be happening? My mom launched out of her seat, so I had to follow. Even at this age, I knew to protect her. At least I would try. A ten-year-old can only offer moral support. We had just moved to town a day earlier. Maybe this wasn't the best decision for our family to make, but, again, the opinion of a ten-year-old doesn't carry much weight. Right now, I needed to be close to Mom as she approached danger.

Only sixty seconds earlier, we were sitting at the stoplight like we'd done hundreds of times before. In an instant, both doors of the car to our left swung open. Two men jumped out while at the same time pulling something from their waistbands. I'd seen something that looked like the things in their hands before. These were guns. The two men walked to the car to our right, pointed inside its windows, and shot both men. I can't describe what it feels like to feel something that has no accompanying words. Whatever that is, I felt it. That's when my mom, who I've learned is afraid of nothing and nobody, jumped out of our car and walked over to the men who'd just been shot. The shooters quickly got back into their vehicles and sped away.

"Are you alright?" Mom asked in the kindest, most caring voice possible during what amounted to a gun fight. Both men moaned in response. One of the men, the one on the passenger's side closest to me, removed his hand exposing a bullet wound in his side. I remember blood gushing out as if someone had opened a high-pressured fire hydrant. While scared and confused, I can remember the calm with which my mom said her next words.

"Thank God they're still breathing."

Right then and there I associated breathing with another chance, an opportunity to get out of a seemingly helpless situation. If you are still breathing, you can make it.

Baltimore

Lead paramedic, JP, yells at the new driver for taking a wrong turn.

"Get us to Mercy Hospital! We're losing him!"

"3! 4! 5!" He's focused on chest compressions, but the victim's vital signs continue to drop. Shot twice in the side, he's bleeding out fast.

"8! 9! 10!" JP won't give up.

Something in this patient reminds him of himself, but he can't place what it is. This kid is so young, too young. Sweat runs into JP's eyes. He grits his teeth. Angrily, he shouts instructions to the victim and a request to the master physician.

"Dammit, don't you die on me, man! God, please help me."

The kid lets out a gut-wrenching scream.

A blast of blood shoots out of the bullet holes.

*Beeeep...*The machine says he's gone.

"He's coding!" JP announces.

The back doors to the ambulance fly open. Nurses and doctors pull the unresponsive young man out of the ambulance. JP reports details to the attending physician.

"Twenty-four-year-old Hispanic male—two gunshots to his right side."

"Is he breathing?" the doctor asks.

"No pulse," JP responds.

"Is he breathing?" the doctor asks again.

"No. He's not," JP responds.

Breath is life; its absence is death.

Afghanistan

Motionless, he lies there. His bodysuit protects against rodents seeking areas of exposed flesh. No sound is made. His only focus is the mission and its success. Precisely when he'll be able to execute his mission is anyone's guess. Preparation is sure; opportunity is not.

The three-day journey to this spot was plagued by inclement weather. He's been sent here by those unwilling to admit he exists, much less admit they sent him. His job is to do a job no one else is qualified to do. This man is a US Navy sniper.

Years of training come down to this moment. Swimming ten miles at 5 am, no problem; walking wet and barefoot with eighty pounds of equipment in tow, no problem. The most difficult task on this day is the same task he has struggled with since day one of boot camp. In order to leverage his expertise and complete the mission, he must properly manage his *breathing*. If he doesn't breathe properly, he will miss his target. If he doesn't breathe properly, he risks giving up his position to highly sensitive equipment designed to identify the slightest error of movement. Despite the hunger, the sweat and the fatigue, he remains still. Did you know every emotion has its own breath? When you cry, your breath cries. When you laugh, your breath pauses to laugh with you. When in pain, your breathing refuses to go its own way. Instead, it labors next to you like a faithful German Shepherd refusing to let his master suffer alone. When sick, your breath is shallower, hampered by reduced optimism. Our bodies must be trained to breathe healthy even when our bodies are sick.

This runs counter to intuition, but pairing human feelings with the easiest breathing patterns must be reversed. Sickness, harm, and depression—just like health and happiness and joy—must be fed or they will die. Starve what you don't want, and feed what you do. Think of the man who has lost all material wealth and stands to lose his family; it is *this* man who must take illogical risks as if he owned the world if he is to have a chance at regaining his piece of it. It is the woman who lies on her bed of recovery after a double mastectomy; it is she who must remember how it feels to breathe in the crisp fall breeze. She must do this if she is to stop slipping further into the ranks of the sick and climb back into the ranks of the well. We must point our bodies in the direction we want to go long before we ever get there.

Turks and Caicos

Sweating underwater is nearly impossible, but somehow, I did it. Well, you just had to be there—deep underwater, alone, afraid. I looked around. No one. No humans that is, only fish, sand, and coral. It seemed each passing second was like a mailman. And his mail was a fifty-pound bag of anxiety dropped right into my lap. I couldn't move. Oh, and the fog of sand in front of me? Well, it thickened into an old western cowboy movie dust cloud. Fear grabbed me totally, but fear is weird. Fear amplifies. Each individual grain of sand looked like a grapefruit. Everything became so clear. Where was my divemaster? I'd lost all sense of direction. When you're this deep, there's no east, west, north or south—there's just open space. Somehow, I remembered his often-repeated instructions. "Control your breathing. You must control your breathing."

As quickly as he had vanished, he reappeared.

"This way, this way!" His hand waved in animated strokes.

My divemaster was signaling which way I needed to head to rejoin the group. Seemingly without delay, my anxiety turned into fun. In the blink of an eye we were swimming with manta rays, feeding them, and watching their every movement. Some of these rays had a two-foot wingspan, some had six, and some were tiny little things. You could see the difference between females and males just by their size. The rays breathe from a hole underneath their bodies. We had squid to feed them. They would breathe in so aggressively that an air vacuum was created, pulling the squid right out of our hands. One manta ray in particular was a bit aggressive. There's always one. This one wanted more than his portion of the food. In Napoleonic gesture, he came up to me and pushed out his chest. This created such suction that all of the squid in my hand went into his mouth. I noticed he was breathing selfishly. For me, this provided a good lesson on the intentionality of breath. The manta ray knew how to get what he wanted, and he used breath to get it. Do you?

Connections. Breath has particular meaning at the bottom of the ocean, but connections still must be made. There is a place underwater called the cleaning station. I first learned about cleaning stations from my client Bonnie White, who is a photographer and teacher. This is where all underwater creatures come to be cleaned throughout the day. The curious thing is this: predator fish that would otherwise eat everything in sight will not harm the defenseless cleaning fish. A cleaner fish picks all the parasites, dead skin cells, and bacteria off of surrounding fish. Fish that are natural enemies of cleaners recognize their place in the ecosystem. Ectoparasites that attach to fish during sleep time can only be removed by the cleaner fish. The cleaner fish will sometimes go into the mouth of a predator fish it is cleaning to make sure it doesn't miss a spot. The underwater experience is a direct metaphor for support. Those who serve have the power to readjust the order of prey beneath the sea.

Pull back from my dive experience and consider a few things in analogy. What is the human cell? What is the nature of a vast ocean? What is hope? The answer is: all are the same. The ocean mirrors the cell, and so does hope itself. We've already learned from Winston how hope is the mitochondria of the human cell, how hope is the eternal, biological intellect of the cell, how hope authorizes the actions of the cell by providing the energy, and the fire it needs to do work that can only be described as a phenomenon beyond comprehension.

When a client sits in front of me, their breathing reveals everything—the dimness in their eyes, cloudiness behind their eyelids. I can see the connection and the disconnection. I can see through the activity of their limbs whether their body is receiving what it needs. I can see in the circulation of their hands and nails whether the blood is being cleaned properly. I can see whether they are using every genomic trigger needed for optimum health. I look at their cheeks and their smile. I look at their frown. I look at their skin and how it feels to the touch. Is it supple or firm? If they are a smoker, I look at how much this has affected their breathing, as it always does. I look how laborious it may be to take in a simple breath and how they try to hide the strain. These are very telling indications.

൪

Take a moment now. Take in a deep breath, count to three in your mind—now breathe out. Now, once more, take a deep breath in for a total of three seconds and breathe out. Breathing calms us down, slows our heart rate, and allows us to have more control over any situation. Anxiety, panic, nervousness, and cold sweats are all amplified when we breathe improperly. In closing, I want you to know that—

"Can you hear me, Ann-Louise?"

"Excuse me?" Ann responded.

"Who is this?"

"Winston."

"We were just talking about you."

"I know. I heard you."

"Ann, you cannot close this chapter without talking about the most important benefit of breathing. You see, I live here in the cell so I see things from an inside-out perspective. You appropriately mentioned how important it is to make the connection between perspectives on weather cells and human cells, but do you know the real reason why you breathe?"

"Well, we thought we did, Winston. We're listening."

"The most profound function of breathing is how it allows the mitochondria to create energy for us. The mitochondria take most of the breath you bring into your body. 'Why does this matter?' you may ask. This is why. Most people take vitamins but totally forget the one vitamin that is most important to the mitochondria. This vitamin is K2. You see, K2, which is the perfect match to vitamin D, adds a super bounce to the mitochondria. Kinda like the difference between eighty-eight octane for your vehicles and eighty-five octane. The performance difference is obvious."

"All we have to do is begin taking K2?"

"That is correct, Ann. Here is a small list of K2 benefits before I go."

Heart disease: K2 reabsorbs calcium plaques

Osteoporosis: K2 lays down a strong bone framework working with vitamin D

Muscular function: K2 enhances contraction, reduces cramps

Mitochondria: K2 works with coenzyme Q10 to grease the flow of electrons through mitochondria so more ATP is made

Nerve pain? Parkinson's? Alzheimer's? Dementia? Vitamin K2 is helping here as mitochondria are resuscitated

Sleep and your mood are better as K2 has a calming influence on the parasympathetic nervous system

"Wow, Winston. Thank you! We will look into K2 immediately and start incorporating it into our diet. This was a lot; ironically, I need to catch my breath."

Breath is the mystery; in it is hidden the secret of life. Breath proves the existence of the life unseen. Breath is audible, and at the same time inaudible. Breath is visible, and at the same time invisible. It is a certain degree of the activity of the breath and the capacity through which it is acting which makes the breath audible. This shows that there exists something of which we are conscious, the source of which no one knows, which is active every moment of the day, on the model of which the mechanism of nature and art are made.

No one can explain whence it came into this mortal body, and no one can say whither it goes when it leaves this body of clay. One

can only say that something living came and kept this mortal body alive and then left it. Man's true self is the part of his being which knows itself to exist, which is conscious of itself. When that self takes breath as its vehicle instead of the body, then it soars upward toward the utmost heights, toward that goal, which is the origin of all beings.

-Hazrat Inayat Khan

CHAPTER TEN
Resilience

You may be one who has walked the path of suffering. You may also know the feeling you get when the feeling returns to your soul. To touch, hear, sense, and notice the nuances of life that disappear when throbs of pain overwhelm each second of your day. And it is this return to normal life that is so precious. Reaching normal can be an extraordinary accomplishment when you've suffered a "dimming" experience. Days when you sense your dreams and aspirations are slipping away. Normal is precious when sickness causes you to negotiate against yourself and against your very will to live. All who have suffered this "dimming" understand all too well what it means. It means if you ever recover, you will never look at life the same again. The question is, do we remember how to get back? Can we provide directions for another? Have we mapped the trail so a wayfarer lost at sea on the dark waters of sickness can also find their way home? What is medicine really if it cannot take us home?

I've cared for hundreds of patients over three decades, but sometimes you remember one. Standing at her bedside, her glassy eyes connected with mine. She mumbled.

"I wanna live."

Chills ran down my arm. I am trained not to engulf myself emotionally in what is clinical. The force of day-to-day activities of those who survive and the force of those who do not can disable a medical professional from doing her duties. You simply cannot be too personally invested. But before I could remember my training, before I could detach, her eyes had already reached beyond my defense. I felt her every word. The sincerity behind every syllable was palatable. With care, I looked around the room. There sat her ninety-one-year-old grandmother holding a white embroidered handkerchief to catch her tears. In silence, her father and mother held hands at the entryway of the room. Her three handsome brothers and one younger sister, all who drove all night in the snow to be there, were depending on me. What they didn't realize is they were depending on my belief system; they were depending on the grade of my hope. Yet, it was the hope in her heart that recognized and agitated the hope in mine. Only six months later, this young woman, once debilitated by a rare liver disease, returned to my office wearing more than a smile and a yellow purse. She handed me a look from those same eyes that said, "Thank you for believing. Thank you for believing I could live." The impact of this patient on my life cannot be fully communicated. If only I could package and mail to each household the feeling of joy, and fulfillment a clinician has when returning a sick family member back to his or her family. There's nothing I desire more than to make her experience everyone's experience. It is my daily passion. It is my answer to, "Why am I here?"

Health must not be a personal success story each and every time. Health must become congregational in nature, after all, sickness is. Sickness has no desire to affect few; its ambition is to touch billions with its evil hand. This is why *support* is key. Support heals. Support protects us from that which seeks to harm. Whether it is a grandmother communicating it is alright to cry, or support in the form of adult siblings, who by their very presence are saying nothing is more important in this world on this day than you. Support then, as a genomic trigger, is irreplaceable to our success in the area of health. Nothing big can be built unless underneath it exists something stronger than what is to be built. Stronger than iron, stronger than steel, and more effective than research and pharmacology is the tender touch of love. It is love that expands our thinking beyond our thinking to find answers hidden under mounds of assumption. For all I've seen that adheres to the tenets of medical science, for all of the patients who died when their physician said they would, I've equally seen anomalies that would be beyond understanding if it weren't for understanding how support from family and from those in charge of your care can cause the impossible to bow in shame.

The medical community has witnessed counterintuitive discoveries causing scientists to revisit previously held assumptions. One example of this is research published on the apolipoprotein E (ApoE) gene. This difficult sounding word can be easily described. The ApoE gene represents proteins that mediate cholesterol metabolism and assist in the crucial activity of transporting fatty aides to the brain, but in industrialized societies, ApoE4 carriers face an up to four-fold higher risk for Alzheimer's disease and other age-related cognitive declines.[12] Research into ApoE4 investigated one question: to what degree does an urban or industrialized environment change the risk profile of this gene? This is a question that author Ben Trumble of Arizona State University's (ASU) School of Human Evolution and Social Change and ASU's Center for Evolution and Medicine both set out to answer, and what they found was eye-opening. Gene mutation takes much of its transformation from the environment in which it lives. Certain environments raise the risk of cognitive disease from the ApoE gene, while inhabitants

---

12 http://www.news-medical.net/news/20161230/Study-examines-how-ApoE-gene-may-function-differently-in-infectious-environment.aspx

of tropical regions with low sanitation practices, actually benefit from ApoE. Cures are relative. Many flu vaccines infuse low doses of the flu itself into patients. Likewise, it seems, according to the research, that environments full of parasites mutate ApoE into an oddly healing salve as if the inhabitants had already been exposed to low doses of the otherwise dangerous gene. The whole of immunity resilience is often built out of many understood and even many mysterious parts. One of the better understood building materials of resilience are the eight genomic triggers.

Support triggers encouragement, encouragement feeds hope, and hope conquers all. The commercial architect has much to add to our discussion of support. Regardless of structural beauty, regardless of the important activities to take place in the building to be erected, if the foundation is insufficient, the building will fall. Weight is a truth teller. Weight uncovers real strength or real weakness. Weight reveals the competence of an architect's design, and without deviation, your health is precisely the same.

Sickness is weight. We are upheld, we are secured, and we are healed in proportion to our foundation. Our support is our medicine. Proper support firms our physical, emotional, and spiritual strengths. And like changes made to San Francisco building codes after one of the nation's worse earthquakes in 1989, support must now be flexible enough to move, bend, and adjust without breaking. Likewise, those around us must be firm in their resolve to get you better. Those around you must be ferocious about the certainty of your recovery. Yet, those around you must operate on wheels. That is to mean their thinking and their willingness to learn must be malleable because sickness is forever evolving. Yesterday's common cold is today's double pneumonia. For your sake, establish proper *support*. Resilience requires it.

In every hospital today, we have pharmaceuticals to numb symptoms of sickness. Morphine warms the neck and arrests the cold touch of chronic pain, but numbing and warming has never healed a body beyond where its fading did not return an even greater pain. We need more. We seek to first prevent the healthy from fading; secondly, we seek to quantify the process of return to wellness if it does fade—to leave crumbs of vitality on the path so others can get back on the path as well. I'm speaking of making health a transferable commodity so that anyone who becomes sick can take the "process pill" and return whole again to their families. By process pill, I mean a repeatable sequence of actions that render favorable, dependable consequences of strength, energy, and vibrancy to anyone who subjects themselves to this process.

A major part of this idea is the eighth and final genomic trigger: *resilience*. Resilience doesn't randomly emerge. Resilient people can, at times, be difficult to describe. This is because resilience has little to do with muscles, intellect, or a boastful disposition. Resilience originates in the intangible. What causes one person to go through a traumatic event and fail to survive, while another person goes through the same or several times worse and seems to find a way to pass the test? Even if you equalize the hand we all are dealt and then place us in the same circumstances, some will make it, and some will not. This is the nature of resilience. It is the man that stands up on the inside. This does not mean we cannot, in our biological context, investigate and surface traits that allows our bodies to manifest physically what usually exist as invisible strength. Biologically, resilience is a creative, cumulative capacity built from the seven genomic triggers we've previously discussed in this writing, including motion and emotion.

It is something we learn in elementary school. An object in motion tends to stay in motion. The converse of this is also true, an object that is still tends to remain still until it is permanently still. Moving is sacred because when we engage an active lifestyle, we are engaging life itself. Far from the vanity of aesthetics, living an adventurous lifestyle of walking, hiking, exercising, yoga, or whatever form of physical activity you select, gives us life. I encourage my patients to attach themselves to a daily activity of physical rigor just as much as they attach themselves to eating or brushing their teeth. To live on purpose means we have to be purposeful in doing those things that keep our minds and limbs animated.

As we unpacked in earlier chapters, our food intake plays an important role in fueling the wellness process. What we place in our bodies either adds years or takes years away—a rather stark decision. Foods that are processed, microwaved, stored in plastic, and canned have all been killed. These are dead foods. Again, these foods have been killed. What do you think is transferred to your system when you eat food that is nutritionally dead? You become nutritionally dead. Conversely, foods that are vibrant, full of color, full of nutrients, vitamins, and minerals build your immunity and your resilience.

☙

If our bodies have no rest, they have no recovery. The increasing pace of society doesn't allow for this to settle in the minds of the working fatigued. Perpetual fatigue has become a badge of honor. Saying "I'm tired" seems to mean "I'm productive." But as a trigger, sleep can be the key to unlock many health benefits that we now reach for pills to accomplish. Sleep is nature's medicine. Although physically dormant, the body is extremely active while we sleep, taking somewhat of an inventory of the day, finding what needs to be repaired, and repairing just that. But this requires quality sleep, and finding quality sleep can be a challenge.

When quality sleep hides from my clients, I ask them one question, "What are you taking to bed with you?" This is a rather unearthing question because many take bad food and bad thoughts into the bedroom and then wonder why they cannot sleep. This includes fast food, greasy foods, and foods that are charcoal-broiled or spicy. There's no wonder so many people wake up again and again during the night. Eating wrong will never allow you to sleep right. Much like an airplane on a landing approach, you have to level out your emotions before bed. Too much volatility, too much turbulence, too much arguing is only going to make for a rough landing.

Lastly, sleep shouldn't always be the result of physical exhaustion any no more than eating should be the result of starvation. Did you know that the lack of sleep causes many heart attacks to take place between four and five o'clock in the morning? This is when the body's cortisol level shifts, changing the heart pattern. Your blood is thicker this time of morning. Now, add this molasses-like blood state to the stress experienced by those who are up at this time watching infomercials—and possibly thinking about a job they hate with a passion—and you have a terribly unhealthy recipe. This persona wakes up and heads into the week sleep deprived with heart irregularities. This is why quality sleep cannot be encouraged enough for the sake of restoration. Restorative sleep means oxygen is being brought into the body, repairing itself all the way down to the mitochondria level. You must sleep, you must sleep well, you must sleep peacefully, and you must sleep often. Resilience needs you to.

Breathing is another ingredient of building our eighth trigger of resilience. More specifically, proper breathing. It is a common place assumption that anyone who is breathing is doing just fine. While there exists a logical premise for this position, our quality of breathing closely affects our day to day quality of life. The notion that living isn't enough permits us to seek methods of living better. Deep inhalation, meaningful exhalation, breathing with peaceful intent, and controlling our emotions by controlling the volatility of each breath allows us to live better. When we live under the scrutiny of our best life, when we submit our daily practices to best practices yet to be attained, all that previously qualified as enough simply is no longer enough. Breathe well; resilience needs you to.

Motivation. What truly causes the human mind, the human spirit to reach further, run longer, try harder, and refuse to give up? In our previous chapter on motivation, we discussed the motivation of a seaman capsized at sea. He found in an otherwise useless bottle of soda, the will to stay alive long enough to leave a final message his family. But must motivation pivot on extreme circumstances of trauma, sickness, and emergency? Greater it is to be motivated as if in extreme circumstance, while going about serene, common, day-to-day activities. That is resilient motivation, a reasoning that turns an intense urgency into a peaceful habit. A sort of imperative reasoning that has been normalized without catalyst. To be motivated for greatness while you are great. To be motivated to be healthy while you are healthy. To be motivated to be fit while at the peak of human fitness. To have an umbrella that doesn't wait until it rains before covering us. Internal, unassisted motivation transforms peak performance into our status quo.

So then, we have our community of triggers: support, motivation, breath, sleep, motion, emotion, and food all working in harmony to create resilience. Resilience is distinguished by its inability to emerge independently. In fact, resilience doesn't start off as such. In its infancy, resilience resembles weakness. We see this when a client crosses the threshold of our clinic tired from seeing six to seven practitioners, tired from anticipating another failure. Their body language speaks. They come to us low, upset, and frustrated with charts and useless data points given by those who took their money and nearly took their hope. But hope, that indistinguishable fire of the soul, always drops regenerable seeds in the soil of our greatest of disappointments. Hope simply refuses to die.

Resilience borrows from motivation and works from the inside out. Resilience is quite ironic. One must have resilience to build resilience, much like one often needs money to make money. That said, resilience must start somewhere, and that somewhere is called a "belief."

The belief in not giving up. The belief that trying something and somebody new can refer new results and new insight. The decision to break from the miserable, mediocre masses. The decision to walk through the door of our clinic and engage with a team that refuses to allow your sickness to continue on our watch. Resilience, or the ability to sustain strength over time without diminution, originates in the mind. This first internal step happens long before resilience builds itself to the point of being observable to the naked eye. Winston would broaden our understanding and say resilience is nourished by memory, even if the memory is not our own. He would say resilience can be synthetically built by remembering how others overcame similar obstacles we now face. Winston would claim hope is so far reaching that even by rehearsing the successes of others, we can build resilience for ourselves, that while our life may not house the evidence to justify such a build, we can build nonetheless. How "little" becomes "much" when placed in the masterful hands of hope. He would remind us to hope, to build, to resist! Resist fear, our ultimate—yet beaten— opponent. Fear is an opponent without teeth. Fear is a fighter that only beats those unwilling to step in the ring. Fear has never won a fight it fought, but has also never lost a fight it wasn't made to fight. Hope cradles the cell; fear trembles and quakes it.

It is in the field of carpentry where we gain another perspective on how to view resilience as our powerful last genomic-trigger. When a carpenter weatherproofs the deck of a ship, what is he really doing? He is replacing the wooden surface with a substance harder than wood. He is making the wood stronger. Interestingly enough, the substance used to harden this wood is relatively worthless by itself. It is the coalescence of the two that creates a hybrid material stronger than the materials alone. Weatherproofing exists because the elements to be endured over the lifetime of the ship—namely, rain, salt, and wind—will present a level of aggression too harsh for wood to sustain its character and its integrity. Wood is fibrous, which makes it strong, but wood is also porous, which makes it weak. The cellulose fibers in wood, a linear organic compound, can also be found in the cell walls of green plants. Ever heard you need more fiber in your diet? This dietary fiber is similar to the "woody" fiber found in wood—yes, the wood on the deck of our ship.

The porous nature of wood, or its ingrained weakness becomes a problem when an agitator gets inside the wood and deteriorates its natural composite. The proofing process allows for wood fibers to remain strong while sealing the porous elements to the point that nothing gets inside to destroy the wood from the inside out.

In other words, integrity is maintained.

Integrity is an interesting word within the context of our discussion. We are familiar with the social connotation referring to the presence, or the lack thereof, of private moral discipline in humans. That said, integrity within the guise of materials has an odd similarity. An individual, and a piece of material both have character. When a piece of wood or steel has no integrity, it is then representing itself beyond the point where it can sustain such reputation. This means the veracity of reputation is only good when unchallenged. For example, a piece of steel without integrity can be falsely relied upon based on the general reputation of steel, only to fail its owner. Such a failure of steel can be catastrophic if it holds together an axel of a car or medical device used in critical procedures.

Back to our carpenter. The weather proofing presents *resilience* as an addendum to the original material, an add-on or a necessary appendix. This hints to us that the original material was not designed to withstand what it is now withstanding, but it also suggests that the original material is a perfect substance to make stronger, that such modification makes it more dependable than any other stronger, yet unmodified material used in its organic state.

Everything strong starts from simpler, weaker origins. Look around you. Everything started from hydrogen building blocks.

One hydrogen atom is the simplest—and lightest—atom in the universe. It is a simple, positively-charged proton bound to one single electron. This joining creates a kind of celestial weatherproofing. The universe then fuses hydrogen atoms together to make harder, stronger, more resilient materials like iron. In fact, when you pick up a slab of iron, you are holding the dense fusion of several weaker substances. We may not think about it in terms of nuclear fusion, but we know how resilience is built every day. We see how a soldier enters the armed forces as an inexperienced, frail young man. But from the rigors of training and the influence of hardened instructors, this soldier arrives at a nearly impenetrable physical and mental peak. He is then ready for war and has the resilience to maintain his *integrity* and character, if ever captured. Preparedness is best built from raw, unprepared parts. Your health is no different.

Very few people peak in resilience while in their natural form. When you respect a person for the resilience she exhibits, what you are really saying is you respect the partnerships she made. You respect the other triggers in her life working together on the inside to present what you see on the outside. What you're really saying is she recognized a particular weakness and joined forces able to give her what's needed to perform at her highest level. What other area of our lives is more important in its ability to resist weakness than the area of our health? Each day when we arise, there are literally billions of bacteria seeking to sicken our physical bodies. Communicable diseases are as plentiful as lint in a washroom. Without proper resistance, a biological waterproofing as it were, we are unable to successfully fight against that which would overcome us, but this takes partnership.

Here's the irony. When we are presented with something good that can help us, we often *resist*. Now, this tendency to resist against that which can help us is an entirely different discussion. Using our premise, the natural question is, "What negative powers are we partnering with that keep us from doing what will benefit us?" This negative, diabolical resistance seems to come out of nowhere and only comes out when something that will help us present itself. It's hard to get your child to eat healthy food, but you don't have to convince her to eat candy.

Negative resistance is the opposite of waterproofing; it is stripping. And this stripping is rather curious because if the same carpenter we spoke of above tried to strip the deck of our ship, he may use a substance called turpentine. Why does this matter? It matters, even if only poetically, because turpentine is a fluid obtained by the distillation of resin obtained from live trees.[13] The same tree that is fibrous and porous has in it the intrinsic ingredients to protect and build, or to strip and tear down, and

---

13 https://en.wikipedia.org/wiki/Turpentine

so do you. You have, in your cells, precisely what you need to be sick and what you need to be well. It's all inside. Even what's wrong with you keeps something that would do you even worse out of you.

So how do we build the resistance in our own bodies to protect us from what is detrimental while allowing entry to what is beneficial? Let's revisit the power of decision.

When you decide to be healthy your resistance is immediately being formed. A decision, in and of itself, sets your cellular directions much like a splint tied to a wayward branch directs its future growth. Your decision is the first building block to block sickness. Decisions combat disease.

A simple search on the term "disease" renders a common-sense explanation that is routinely ignored. By definition, disease is *a disorder of structure or function in a human, animal, or plant, especially one that produces specific signs or symptoms or that affects a specific location and is not simply a direct result of physical injury—a particular quality, habit, or disposition regarded as adversely affecting a person or group of people.*[14] Highlighted in this academic definition is "disorder" and "habit." Without a closer look at disease, one can easily assume it to be an unavoidable genetic disorder, and at times it is precisely

---

14 https://www.google.com/#q=disease+definition

that. But often, disease implies the need for a physiological restructure. Disease is a function of our environment, our team, our personal influences. For a moment, let's objectify the body and make it the focus of an analogy of corporate restructuring. Restructuring is a business management term which means to reorganize the legal, ownership, and operational components of a corporation. The sole purpose of a reorganization is to make the targeted company more profitable. Your body is no different, except profit for the body isn't money; it is health and wellness.

The sole purpose of leveraging all aspects of hope by way of support, motivation, breathe, sleep, motion, emotion, food, and resilience is to restructure your physical body and elevate its "health profit"—to turn your health from a liability to an asset. Decisions forcibly take us where we need to go to do what we've decided to do. But be mindful; deciding to be healthy is mockery if we don't take the next step to execute our resolve. Resilience, then, is first a decision to become resilient. It is a process of leveraging everything inside of you, good and bad. We must join forces with clinicians who are committed to sentencing us to living a higher standard of life, clinicians who have the knowledge and know-how we do not have. Resilience ultimately is a partnership. Often, my clients will walk through the door and ask one efficient question, "Will I find the answers here?" What a wonderful question, a question that's clear and directed only comes from not asking enough clear and directed questions previously. And to that client question, I answer an emphatic, "Yes!"

There are a few additional points about disease deserving of expression. We covered the medical definition, but there is another trait of disease, and that is its delusional leanings. A person can think their way into sickness and worry their way into disease even though they may have no physiological propensity or remarkable test results associated with their newly acquired ailment. So where does it come from? How does this happen? Sometimes it is stress; that is, stress of being uncertain as to the future of you or your family. This can be seen when studying telomeres.

Telomeres, which are regions at the end of your chromosomes that protect them from degradation, provide insight into how people get sick. Telomeres also give us a peak into why sickness is rarely random. In a 2012 presentation given by Dr. Gabor Mate, physician and psychoneuroimmunology researcher studying causation between stress and disease, he summarizes one medical study quantifying how individuals who are under great stress, such as mothers with chronically ill children, have a telomere age that is sometimes ten years older than the mother's actual age.[15] This means, in essence, the mother is at risk of dying ten years sooner than she should. Dr. Mate also makes the point that the inability of a caregiver to say "no" is a direct, contributing factor to the future sickness suffered by the caregiver, and that the greatest challenge of the caregiver is they routinely don't care for themselves. They are deluded into thinking they can do everything for everyone. If you don't know how to say no, your body will say it for you. Having a medical and personal team that understands this is vital. Resilience requires rest; resilience requires "no."

In his book, *The Disease Delusion,* author and functional medicine pioneer, Dr. Jeffrey Bland, details the danger of treating symptoms of a disease rather than digging to excavate the true cause of the sickness. Delusion gives us reason to be encouraged. It just depends on how you look at it. If one can delude themselves into sickness—if one can transform a mental state into physical manifestation—it then follows that one can "delude" or believe themselves out of that physical manifestation back into health. The sick can believe themselves back healthy. A part of our ten-year fight plan designed for client wellness is being fit in our thinking. It takes skills, strategies, and new habits to move from weak to strong, from sickly to healthy. It's not a quick fix but a

---

15 https://www.youtube.com/watch?v=Qf92I7FPyKo

long-term strategy to get you there. The fight plan involves life planning, goal planning, and a deep look and strategy into the cellular world within us. We must boost cell health. We must enhance your cell environment and provide your cells what they need to multiply, divide, repair, and replace. We build a new you inside of you. It is a process, but in the end, we've moved your ending far out of sight so you can enjoy each moment of every day with those who mean the most.

༺

On the Trail
by Ann-Louise Johnson

Footsteps never thunder in a forest. They tread lightly like the deer, who jump through the rainbows on the trail without bending a ray, then go silently on.

Footsteps never thunder in a forest.

The thundering is in your heart. I can feel it beating from here. Look within and ask is your heart beating with adventure or fear? Adventure and fear are the same, just opposite sides of the same face.

One smile is climbing the trail, determined, walking tall like the forest trees pruned by strong winds. The other sweats, standing still, shaking, feeling the fear that isn't. Both hearts beat like thunder. One pulses; one quakes.

How can these two, adventure and fear, walk the forest of life together? How can they dance over the life trails like a deer? How can they stand in a rainbow without bending a ray? How can adventure and fear meet and walk together, tall, and strong?

Adventure and fear meet on the trail of the wild heart strong. They face fear and adventure together. They face fear boldly and watch it melt into the face of adventure.

If your heart fears the rose thorns cutting your feet, let me pull them out with a rose petal and soften your trail walk.

If your heart fears the rocks that bruise and scar, it's courage that gives story to each bruise and scar.

If your heart fears to walk on the forest path before sunrise, catching a glimpse of the doe with fawns waking. I thirst for this... Walk with me.

If your heart fears and quakes with the wild wind blowing, touch a tree bending in the wind and know that's why it stands tall.

If your heart fears from the wolf who watches, chilling bones as the hair stands taller on your arms, taller than you, stooping. Listen. She's protecting, not watching you.

These are the way of the wild heart strong, the wild heart answering the call of the wild with each footstep explored.

A strong heart discovering God's breath in the living vein of a leaf, inspiring heart, sharing and releasing newfound strength within, painting a new you, painting heart, brushing fear with a new face, etched in each footstep on the trail, courageous heart. Footsteps never thunder in a forest but the heart thunders with adventure, thundering heart. So walk the wild heart strong trail with me. Go onto the trail. Be wild heart strong with a wild heart.

Go strong. Go inspire. Go painting. Go encourage. Go thundering. Go now. Go. Because the courage and the strength await you as you move into the forest, as you move into strength, as you move into the life, gaining that resilience, looking at that resilience, looking at those triggers of resilience that have escaped you and now can be recaptured in your health story.

CHAPTER ELEVEN
Highways of the Mind and the Cell

There is nothing more emotional than a memory. Nothing reaches deeper into our soul than a childhood memory or pictures of a family member whose hand we will never again hold. As so, it is a special violation to have such memories ripped from us or from those we love. Dementia and Alzheimer's—both are thieves, biological criminals. We suffer at their diabolical hands. Adult children understand this best, especially, when they look into the eyes of their parents only to see a blank stare looking back. The need to repeat the same details is a constant stress, a constant reminder of the relentless nature of cognitive disease. It hurts. Today, there's an exciting new breed of researchers who refuse to allow such pain to continue unchallenged. There is a new thinking. We now know the answers that we have been seeking, have actually been seeking us.

Take a moment. Look around you. While sitting in traffic, pause from being frustrated and notice what mankind has done.

There isn't a place on earth we can't access. We have created millions of miles of roads and highways taking us to the most remote places on the planet. Not only that, but if a road, a bridge, or highway fails from natural or unnatural causes, we, through the power of civil engineering, can immediately create detours to give us the same access as before. Where did these ideas come from? How did mankind figure out the process by which we can rebuild and reestablish old connections in a new way? This skill, this knowledge, did not originate in school; it originated in the mind. Think of this differently. The mind will often take the form of creativity and create structures outside of itself to teach us what can be done inside of itself. The mind is so ingenious that even if we cannot grasp the width and breadth of its intrinsic ability, it can simplify its ability and present its capacity in a different form. The form of roads and highways allows us to drive every day on what with think are pathways of transportation, when we are actually traversing on neurons the mind has materialized in a different form to speak to us in a language we understand.

And it is this, it is the thinking mind operating within the mind itself that possesses an insatiable appetite, an appetite for bettering our understanding of what the mind can actually do. The mind reaches outside of us and creates things that reveal the unlimited reach inside of us all. You are a walking trove of healing, a bottomless depth of answers. You are a fractal piece of humanity, while also an unremitting layer of unrepeated creativity. Every day, the cells in you sing a new song of joy, clarity, abundance, and bold adventure. You are a life form that refuses to live someone else's life. Whatever has killed someone else does not have to kill you. Death has no memory, life has no ceiling.

Therefore, let not your life be defined by time spent on immature endeavors. I speak of an immaturity not from adolescent age, but adolescent thinking. Thinking that is resistance to that which you've never seen or heard as if seeing and hearing are the basis of truth. Your body is a creation, and your mind its central headquarters for accomplishing the impossible. The slow, but ever-increasing spiral of sickness that often accompanies older age, the loss of memory, and the reduced activity of our limbs does not have to be your destiny. It is your choice, and you can choose differently. Nature, of which we are the crowning achievement, has gone before us and set the different pace. Nature is nauseated by uniformity. Nature is diverse and eats fruit from the tree called impossible daily. Nature wakes up each day knowing that in it is everything needed to shine brighter, water more plentifully, and be warmer and more inviting than any day ever before. Nature pays no attention to age nor the decaying hand of father time. You must approach each day in the same manner. You are nature.

So then if we, the scientist, the athlete, the executive, and the day laborer, are to access longer life and uncover the simple solutions to its complexities—especially the complexity of cognitive decay—this will not happen in research laboratories alone. A longer, more quality life will happen when we, with a different eye, look around us and see the clues our minds have left right in front of us. We are magnificent beings able to do the impossible with ease.

☙

It is Tuesday, 7 pm, and like each Tuesday for the last three years, Donna enters the room, pulls back the curtains, and allows the sun's angled rays to warm the petite space. Everything is just the way she left it. The television plays *Law and Order*, the UNO cards lay tired on the table. Pulling the door to a soft close, Donna sits down placing her hand on one foot peeking out from the covers. She then reaches into her purse to repaint the weakening pink polish. The elderly woman awakens.

"Mom, mom?"

The eyes of the beautiful silver-haired woman search the room for the voice she heard. Finding the face, she smiles, but quickly, her smile fades away.

"Mom — it's me, Donna, your daughter."

Donna, who is a spitting image of the older woman she now visits three times per week, is holding back tears. Her mother looks deep into her eyes, and with sincerity asks the question she asks each time they speak.

"What's your name baby?"

Donna moves closer to her mother's face. Her mom then places the back of her hand on Donna's cheek, and with a slow touch begins her routine compliments.

"You're such a beautiful young lady; your mom must be proud." By now, Donna is shaking with emotions. Her eyes dam with tears.

"But mom, it's me Donna. You are my mother. I'm little Donna." The woman smiles out of courtesy, not out of understanding. Donna's mother has dementia, the disease that kills while keeping you alive. Dementia and its older brother, Alzheimer's, kills the memories of years, people, and events gone by. The cognitive decay associated with such illnesses leaves loved ones as a mere shell of their previous selves. The question is: does it have to be this way? Surprisingly, it does not. The natural question that follows is this: why do millions continue to suffer this debilitating illness if these diseases are curable? Even if with a jaded eye, the public looks at the medical establishment and claims there is a profitable reason for allowing this class of disease to continue, you still must answer one unrelenting question. The researcher, the doctor, the scientist, and the president of your health care company have parents. Many of them have parents who are suffering from some sort of cognitive disease, so then, it is not easily explained why answers exist that aren't being used to reduce the pain of the wealthy medical professional, much less the average citizen. In reality, the medical professional and the regular citizen alike often suffer from thinking gaps. Yes, even the most educated, brilliant minds we have experience thinking gaps that prevent the masses from accessing answers the mind has laid prostrate right in front of us since the beginning of time.

Another challenge of seeing obvious answers nature has placed around us is *support*. Support is the foundational trigger for all genomic triggers. Finding support for new curative approaches to dementia and Alzheimer's has the same obstacles as anything else where the problem has become an accepted norm. When problems become accepted norms, answers can be viewed as worse problems than the problem itself. When one normalizes suffering, the absence of suffering can be an odd inconvenience. Pain has its own pathology of comfort. Sometimes, you have to fight for what is good and right. Regardless of the potential benefit to mankind, nothing of any value is given, and even good must be forcibly accomplished.

So what would happen if we changed our thinking about cognitive disease and found the support to make substantive change across the world? What could we do if we could proactively increase our brains cells? More brain cells mean less nursing home time. Long-term health insurance could become a thing of the of the past. Without it, our geriatric population decays down to nothing. Our elderly are stripped of their memory and their abilities to do basic tasks. They lose their words. They are ostracized, and in return, they ostracize themselves. Their world gets smaller and smaller. No more grocery store visits because they don't seem to recognize neighbors who recognize them. They eat their way deeper into dementia by feasting on cookies and crackers all day. It is only when you lower your carbohydrate intake that cognitive performance increases. Food has no neutral gear. Food heals, or food kills.

Pressure to change this reality will not happen until cognitive disease becomes the health care civil rights issue of the twenty-first century. This requires a champion of the cognitive cause to commit to raising awareness. We've seen the life cycle before. We know how it goes. A zealous, outspoken individual organizes a committee. The committee organizes rallies. Rallies garner media attention. Media attention gets the general public to pay attention. When the general public is interested, legislators become interested. Bills are proposed, votes are taken and eventually, cognitive decline initiatives are on everyone's radar. It starts with one person. It starts with you. It starts with me.

Historically, public outcry has been the most productive form of "protest" able to reverse the diabolical comfort and acceptance of pain and suffering. Public outcry can right the philosophical ship. Public outcry can transform individual concerns into public law. Whether the concern is equality or medicine, the general public is the one mechanism that can overcome political and monetary power by itself becoming a larger political and monetary power. This requires health care to become a community issue instead of a personal matter. Do you know why you don't have to worry about finding your own power source for your home? Do you know why you don't have to worry about hiring your own local paramedic or digging water wells for your morning shower? It is because somewhere along the way, communities discovered it was better for all residents to benefit from a central power grid and a central water and sewer system. The community decided it was better for us all to have an emergency system that services the entire county rather than the independent concept of, "every man for himself." This does not articulate a perspective based on political socialism, but rather pragmatic intelligence. So what about health care? What if we viewed health care as a community problem, one the entire community must solve? Again, just like the highway and roads around us that have been physicalized, the mind has created unrelated examples of mutually beneficial systems in order for us expand this approach to fields that could use it most. Do we, as a society, have the courage to apply these lessons to uncomfortable areas?

Speaking of champions of the cause, we may already have one. Just ask Dale Bredesen, MD about the difficulty of getting broad support even for discoveries that benefit everybody. Reversing cognitive decline—his area of research—is far from a new concept. It is often the case that disease, especially severely degenerative diseases, when left alone will "remain alive" due to metabolic processes. That is to say, organisms suffering from disease negatively benefit from the normal functioning processes of the human body. Both anything that is sick and anything that is well grow. To live, tumors must be fed, they must respire, they must obtain and release energy, and they must produce carbon dioxide. Processes associated with healthy people and healthy organisms are the same processes that keeps disease alive. This is where therapeutic approaches become vital.

Dr. Bredesen of The Buck Institute for Research on Aging has presented quantifiable results to the possible treatment and great reduction of cognitive decline due to neurological disease. The increasingly pressing need for treatment to Alzheimer's is articulated by its aggressive rise within the current world population. According to Bredesen, as of September 2014, 5.4 million people in the US suffers from Alzheimer's, 30 million people globally. By 2015, these numbers are projected to rise to 13 million and 160 million respectively, a move that threatens to bankrupt the current Medicare system. By the time you read this, those numbers would have increased. A therapeutic approach to neurological decline breaks from current monotherapeutic practices that have failed to reduce the number contracting this class of disease. Therapeutic approaches also mitigate the "metabolic help" cognitive decline disease receives by casting a larger net of treatment instead of depending too heavily on targeting approaches. The Bredesen Protocol addresses this.

The protocol leverages a comprehensive personalized system designed to improve cognition and reverse cognitive decline of due to early Alzheimer's disease.[16] There are scientific and anecdotal encouragements for this approach. Oddly enough, combination therapies have been central to increasing success in treating cardiovascular disease, cancer, HIV, and other serious illnesses. It is the area of cognitive design that is far behind in using the same approach used to solved other diseases. We must close the gap and close the gap now. People are waiting far too long ago address cognitive disease. You must be ten years ahead of its arrival. Once it arrives, it is often too late.

In case study after case study, the protocol has accomplished reversing neurological illness in patients, returning them back to their families, and send them back to work with their mental capacity in tact. In one case, a sixty-seven-year-old patient was

---

[16] https://www.mpi-cognition.com/protocol-overview/

experiencing progressive cognitive decline. Her mother died with dementia, and now she found herself not being able to navigate the freeway. She couldn't remember anything she read or even recall four-digit numbers. After being treated with Metabolic Enhancement for NeuroDegeneration (MEND) her decline reversed, and she was able to return to work full-time with a memory that was better than it had been in decades.[17] And this is among numerous case studies with similar, if not better results. Your mind can do so much more than you are requiring it to do. Your mind is a builder. Your mind is a master engineer. Your mind is a construction site. Your mind is your future today.

☙

It is Tuesday again, 7 pm. It's been three years since Donna first visited her mother in the nursing home. Today is different. On this day, Donna begins to notice something. Walking outside, she can't remember where she parked. Dismissing this as fatigue, Donna doesn't pay it much mind. But two weeks later, she notices she can't remember her PIN at the ATM. A month later, Donna doesn't show up for a client meeting, an omission she hasn't made in twenty years. This is when Donna makes a critical decision that changes the trajectory of her life.

---

17 https://www.mpi-cognition.com/protocol-overview/

There is an active retirement community where Donna lives, a community that believes we can grow our neurons and grow our brains. Donna needed support. She was confused and scared. She met with a certified functional medicine practitioner who explained their in-house cognitive growth protocol. The community has a group that works with day patients on healing the first indications of a wounded cognition. They believe in starting the healing before sickness takes root. Donna also met a chef who manages a personalized menu and appropriate accompanying supplements. There is music training and a physical fitness regimen targeted for the mind. This assisted living day center is designed not to take care of the sick, but to prevent the sickness by noticing cognitive decline at its onset and immediately halting its expansion. And it is this network of support, a team of professionals synchronized in their respective efforts to return each patient to their best life form, that prevents Donna from following her mother into the dark cave of memory loss. Donna exercises each day now with the knowledge that exercise grows the brain because of increased oxygen. She is no longer confused, nor is she scared.

The lifestyle engaged for Donna to overcome her genetic bending toward memory loss has not only reversed her course, but given her a new, sustainable healthy life. In the community center, Donna learned that certain foods boost brain health, while other foods support mitochondria doubling. This could not have been done at home all by herself. The healing nature of food is only information unless you have a team exampling its power around you each day. With this new knowledge in hand, she became motivated. The surrounding support team gave her hope.

Emotionally, she has learned to manage stress by using natures oldest supplement, laughter—especially the belly kind! It was difficult to include exercise in her normal routine, but she quickly noticed how exercise was reshaping her life, tightening her body, increasing metabolism, doubling her mitochondria, and simply giving her more strength. Donna was taught to breathe deeply and more regularly. She learned the importance of sleep by reading applicable research materials. She discovered certain "sleep genes" that responded well to regular sleep which caused her to protect her sleep with a structured early bedtime. Sleep is when we put our body in the shop for repairs. In the end, Donna learned that we must, through our eight genomic triggers, provide our cells what they need to multiply, divide, repair, and replace. It is a process, but you can build a new you inside of you. You can change course and be healthier than ever, but you have to have the right support. You must place yourself in an environment with people who want nothing more than for you to win.

One year after her first visit to the community health center, Donna is leaving her mother's nursing home. She sits at the stoplight in deep thought. Traffic is heavy, but Donna feels an odd calm as she takes a moment to notice the spaghetti-like highways crossing each other in an orchestra of organized confusion. No road runs into the other. Each highway moves a different group of cars in a different direction.

"What a masterpiece." Donna says to herself.

"The mind is such a masterpiece."

CHAPTER TWELVE
The DNA of Hope Process

Handsome and physically fit, Timothy Jeffries walked into my office lobby. He took strong steps. His smile was broad, bright, and welcoming. For some reason, I even remember the change softly touching in his pocket chiming random melodies. Mr. Jeffries looked like a healthy, middle-aged man. No darkness in his eyes. His countenance was free from the remnants of hard living that so often follows a person long after their hard-living days are over. His cartage? An ideal weight for his age and stature. Still, he was in pain.

But sometimes sickness hides behind the wall of our emotions. Sometimes sickness hides deep in the wet caves of disappointment, fatigue, and unfulfilled dreams. While others may see the picture of physical health, there is often sickness in places no one can see or touch. In places no one knows exist. When I walked into the examining room, his eyes were looking upward.

"Powerful." He said.

I glanced upward to see the posting that so often catches the attention of our patients.

The Hope Creed

Hope is bigger than the tallest mountain,
Yet small enough to live in me.
Able to calm the winds of sickness, and still the raging sea.
Hope is a pillar of sunshine, and no respecter of man.
Hope is fair, speaki ng childlike terms everyone can understand.
Hope reaches eight times into the soul.
Hope warms blood once cold.
And though others surrender to disease misunderstood,
Hope never gives ground, seeking the best over the good.
Hope will never touch a path previously trod,
But creates for the next traveler a manageable sod.
I believe in Hope.
Hope disrupts, and finishes the race.
Hope is committed, and maintains feverish pace.
So if in a pit you ever lay, your brow bloodied from the fight of the day,
Hope will carry your load when you've done all you can do.
And at the moment you want to give up on Hope,
Hope will never give up on you.

"I felt that."

"What do you mean?" I asked.

"Those words. They reach deep into you. They open a portal." He said.

"Hope is the most powerful force on earth. Hope can force into you or force out of you whatever it is you hope for. Nature had physicians before society had medical schools. Nature had medicine, surgeries, and recovery processes long before these practices were ever formalized. It was called hope."

Something sparked in his eyes. We began to talk. Mr. Jeffries had recently completed The DNA of Hope Workbook called *Operation Hope*. While many ailments found therein did not apply to him, he did sense there was something missing in his life. He sensed that the cellular connection to hope, and hope's power to rewrite our lives had pulled back a corner on feelings he'd had for a while but could not verifiably articulate. After looking at his chart, I began the all-important intake process. I need to find out his life's goal. I have to find out what matters most to him because this and this alone determines where we go from here and the best treatment route to travel. Over the next hour and a half, we talked. With hands clinched against his knees, Mr. Jeffries began with questions.

"How do you find out what's wrong when all your medical tests say nothing is wrong? How do you say what you feel when what you feel has no words? I've felt for years that something wasn't right, but that's about as descriptive as I can be. I don't quite know what else to call it. I simply don't know. If it were a pain, I could point to it. But it's a bit more complicated than that."

He sat there shaking his head fearing I wouldn't understand, but I understood. His unspoken words were loud and clear to me. I empathize with the challenge of expressing that which is so difficult to express. The very thing that is nearly impossible to articulate can haunt a person all of their days until that "something" is harnessed. That is, until you have a name and then work to prevent that thing from running wild through the emotional forest of your life. I decided to press further and cause him to dig just a little deeper.

"You may not know what to call what you feel, but let me ask you this." I said.

"Name three things you would change if you could right now." Without hesitation, he answered.

"First, I'd sleep better, then I'd eat better, and then I'd remove toxic people from around me—not sure if I'd do those things in exactly that order, but I need all those things done.

"Got it." I said.

I now had the three triggers we immediately had to deal with in order to turn his invisible liabilities into tangible assets. We had to discuss sleep, food, and support. These three concentrations turn the light on in the darkness. And because he identified what means the most to him, the process will be more meaningful, which in turn means the process will be more effective.

I began by explaining how our sleep genes influence our

behavior and ultimately, our health. I took time to unpackage the fact that sleep is a course in miracles. While sleeping, the organized complications of your cellular, circulatory, respiratory, and muscular systems operate without conscious command. Only in sleep does our body enjoy comprehensive access to its beauty, performance, and intrinsic genius. In the sleep state, our body heals itself. It refuels our immunity to face dangers awaiting once we awake. Sleeping is the body's tool shed. We are fixed while we sleep.

I then explained how our sleep genes influence our general optimism towards life itself. Sleep genes determine our emotion and our motion. Whatever is hiding becomes more evident the better we sleep. Sleep fishes out what hides beneath. The more fatigued we are, the less we are able to articulate the emotions that have been muddled by being so tired all the time. We also must never forget how much our sleep genes equally need light and darkness. We satisfy the light requirement in the sun of the day; therefore, we need to sleep in total darkness to encourage the full expression of the genes that only reach their peak performance in such an environment.

Mr. Jeffries jumped in.

"I cannot remember the last time I got a good night's rest. Probably been years. Tried changing my eating habits from late at night to cutting food off in the early evening. Even tried to go to sleep earlier than I usually do. Nothing seemed to help. I just lay in bed looking at the ceiling most of the time thinking about nothing and everything."

I didn't want to critique every point—at least not at this point in the conversation—so I paused and continued with other questions.

"How is your food prepared?" I asked.

His smile quickly grew into a laugh.

"Wish I could eat better." He said.

"Funny thing is, I know what I should eat. I even know how best to prepare it. I just don't have the time or I don't make the time do it right. Always running, always working, always something else I have to tend to. So, I eat out a lot, and I do mean a lot!"

"Understood." I responded.

"Have you ever heard of *Heal with a Meal©*", I asked.

"Well, I heard about it in your workbook, but I need to activate it in my day-to-day life."

"We'll get it done." I continued.

"From reading the workbook, you understand why food that heals starts with proper preparation. This means we must get back to cooking in our homes and leave restaurant food alone. For many people this isn't a possibility, which is where healing one meal at a time comes in. Just one meal a day or one meal a week, whatever frequency your life can sustain. Then build to one meal every other day, etcetera. Slowly, you'll notice the difference in your body. Food transformation also requires we stop using microwave. If there is one thing we can do today that will immediately help us tomorrow it is stop microwave cooking. Let me tell you why: microwaves create dead food. Microwaves must kill your food to heat it as fast as you want it. And for this speed, you give up life. Have you ever noticed how food tastes differently almost like plastic after being microwaved? To cook your food in seconds requires an electron pinging and cycling more than two billion times per second. Don't you think the electrons lose energy? When you eat the electrons, they are tired, and then they make you tired. Cut back on your microwaved foods and you'll begin to have more energy. Another thing to keep in mind is limiting your use of refrigerated foods. Why? Because refrigerated foods grow certain bacteria that can become a springboard for allergies and other sensitivities."

"Well, like I said, I eat out a lot, I use microwaves a lot, and everything in my home is refrigerated." Mr. Jeffries said sheepishly.

"Gonna cut down on the microwave."

"Great start!" I responded and encouraged him further.

"Whether you know it or not, you have just entered into the process that leads to feeling better, thinking better and ultimately, living better. Let me discuss support for a moment, and then we're gonna go back through with more specifics on all three—

support, food, and sleep. If I asked you to show me pictures of your closest friends, the ones with whom you spend the most time do you know what I can predict?

"What's that?"

"I can predict that you look like them and they look similar to you. You see, we look more like friends than we look like family because we eat with friends more often. Sociology is a phenotype influencer. That simply means by beholding, we become changed. Humans, over time, begin to physically manifest the influences around them. Smokers tend to encourage smoking, drinkers encourage drinkers and those with bad eating habits encourage other to also have bad eating habits. Peer pressure isn't just an adolescent phenomenon. Food is that influential. Food represents the power we put in us, or better said, it represents how we increase or decrease the power that's already in us."

"This explains a lot."

"You mentioned support. It's so interesting how we human beings understand certain concepts in their fullness in every area except the area of our personal lives. We understand how a bridge can never stand without proper support. We understand that what is underneath a thing must be stronger than that which is on top of it. Support in our personal lives is exactly that, but it is even more. You see, those in your circle of support not only "hold you up" (healthy) or "break you down" (toxic) but this group of friends and associates represents your personal jury, a jury that renders perpetual verdicts. Your personal jury is in the business of convicting you each and every day. You will either be convicted to elevate your standard of living, eating, and thinking, or be convicted to reinforce sickness, toxic relationships, and destructive behaviors. That represents the personnel in your life, but you also must have cellular support and mitochondrial support within you. This represents support at the nutrient levels."

Just then, footsteps could be heard quickly approaching the examination room door, and a nurse knocked frantically. She opened the door.

"Mr. Crumpton in room number two is not responding!"

I rushed to the room to see two feet laying across the threshold pointing upward.

"Call 911." I told my nurse.

After checking his vitals, Mr. Crumpton was still alive. Immediately, we began CPR. My head nurse held his hand on my right and my second nurse held his left hand. I straddled his body. We were determined to keep him alive until the ambulance could get him to the hospital. Mr. Crumpton had just become a patient the previous week, and this was to be our very first intake meeting. From an initial conversation, it was obvious he was an incredibly nice man who had been suffering from obesity. He came to us to get control of his diet, thus control of his weight.

"Get his wife on the phone." I yelled out in between compression counts. When you've been in medicine awhile, you can just feel when something is bad. This was bad. For him to come through this event was going to take more than traditional medicine. I felt death approaching. You can feel it coming, but I also knew there's one thing death can't defeat: hope.

"Lisa Crumpton's on the phone!" My nurse shouted out from the lobby.

"Bring the phone back here." I ordered.

"Ambulance is three mins out." My nurse updated me.

"Tell his wife what's going on, then put her on speaker."

I continued compressions, and slowly, I noticed color returning to his body.

"What's going on?" Mr. Jeffries had stepped outside his examination room after hearing all the commotion.

"Patient emergency in another room, sir," one of the nurses swiftly explained.

"Lisa, help is on the way. Your husband is breathing. Just call his name, no panic, just call his name as we work on him." I told his wife. Sirens could now be heard closing in on our offices. Through her tears, Lisa began to call her husband's name.

"James...James, I love you. Can you hear me, James? It's going to be okay. Help is on the way. Help is on the way."

He was still in trouble, but his vital signs began to stabilize.

"James, me and the kids are on our way to meet you at the hospital. Hold on, James. Hold on, honey. I love you."

Paramedics could be heard rustling through the front door.

"Where is he?" They asked.

"Room number two, this way!"

By this time, Mr. Jeffries was standing in the hallway affixed to what was going on. He was trying to manage what he was feeling from what he was seeing. Less than five minutes later, Mr. Crumpton had been stabilized by the paramedics and rushed to the hospital. It took about fifteen mins for the energy in the office to calm down. I cleaned up and decided to finish seeing patients. I waked back into Mr. Jeffries room. I wasn't even a foot in the door before he started talking.

"That was amazing. Is he gonna make it?" Mr. Jeffries asked.

"We hope so. We sincerely hope so."

"I know medical professionals are serious about helping someone in need, but you guys looked like a whole different level. It was like a caring event."

"I've never heard it put that way, but you're right. We care."

"Why did you put his wife on the phone? I would have thought that may have been a bad idea—that things may have gotten too emotional for you to remain focused."

"Good question, and it kind of centers around our mission here as functional practitioners—we believe deeply in hope. First, let me say I know Lisa. She is the one who referred James to us. She and her husband have a great, loving, supportive relationship. Knowing this information, I knew that her voice would birth hope in him even if he could not respond; hope would still do its job. Hearing is one of the last things to go when a person has a remarkable medical event. What a person hears is critically important. This is also why we made discretionary physical touch with the patient by holding his hands. It is our Siamese approach to treating and healing sickness, even if the sickness is just depression and fear.

"Siamese approach?" Mr. Jeffries asked.

"Yes. This is the best way to describe how inextricably tied together a person is to whom they are connected. You breathe the breath of your friends, family, and other close associates. Their sight is your sight, their health is your health. In other words, Joan Baez was correct, 'No man is an island, no man stands alone. Each man's joy is joy to me, each man's grief is my own.' When we have this kind of support, we lean toward being better, getting better, living better. Is it a perfect art and science? No. It is not, but very few things are. The Siamese approach simply puts the odds in your favor, the odds of good health."

"I want support, I need support. Can you help me develop a team?"

"That's precisely what I do. I'm so glad you asked for one

before I suggested one for you! Your team can be a small or large, but you need people around you who believe in you. You need people outside of you who believe in you, people who are going to take a 'failure is not an option' approach toward whatever it is you want and need out of life. You make a list of those who care for you enough to take action for you, and I'll meet with them to make sure everyone is on the same page. We must care, but then we must coordinate.

"We give your team parameters in which operate, and jobs to accomplish. For example, you'll have a laughing team, yep, a laughing team. Sounds funny already, doesn't it? You'll have a cooking team, a shopping team, and a reading team. We put together a conglomerate of care with its sole purpose of getting you where you want and need to be. Together, we help you live stronger and longer. There's also an additional benefit to the team approach. You see, team members eventually become team members for other team members. In other words, the support team developed for you becomes support for others in need. It's the gift that keeps on giving. Word spreads, and those with experience in getting you where you need to be become experts at helping others do the same.

"Teams are magical gap-fillers. Whether a patient has diabetes or is an athlete on her way to the Olympics, everyone seeks peak performance, but no one has everything they need for this peak performance. No one person has all the talents, gifts, or insight needed. Peak performance requires peak ability, which is an offspring of peak capacity. Here's the thing—it doesn't have to be your ability or your capacity. We put together a group of talented, gifted, committed people who together make up for what you may not have personally, but we have collectively. The end result is you win with us as if you won alone."

"I just saw an example of that a few minutes ago with the gentleman."

"Exactly." I said.

"Now, let's get you a plan you can follow to address the three items you told me you want to correct in your life."

"Let's do it!" Mr. Jeffries said emphatically.

"Grab a pen and pad. I don't want to print this out for you; I want you to write everything down in your own handwriting. Your plan must be written down and organized by you!"

Over the next two hours, we talked. Mr. Jeffries dictated his plan in his own handwriting. The DNA of Hope process has to be simple enough for patients to follow, yet gradually intense enough to require more out of the patient as the weeks and months go by as they see the tangible results of eating better, sleeping better, and having a better support system. This is what Mr. Jeffries wrote down. What will your plan look like?

The Health Blueprint for Timothy Jeffries

Start date
Today

Our Aim
Reduce poor sleeping habits
Reduce intake of prepackaged food
Eliminate toxic relationships
Be the best father to my grown children
Start dating again and find a woman who loves life as much as I do

Human Tools
Health care team / My support team

Scientific Tools
Opus 23 software. Functional lab tests

Personal commitment:
I, Timothy Jeffries, am the seed and the sower. I have the capacity to create and live. I have to capacity to destroy and die. Both rest in my hands. I commit my life from this day forward to one lived with meaning and thoughtful direction. I commit my days to extending my days, to embracing the nutritional harmony nature has laid up for me. I commit to digest only that which gives me life, and divest myself of all that does not. With sobriety of mind, body, and soul, I commit myself to a better me, to an adventurous future and a life of indescribable joy.

[My Next Steps]
Protect my sleep - be in bed by 10 pm at least 5 of 7 days per week.
Start *Heal with a Meal*© once per week.
Less prepackaged food. Stove instead of microwave.
Speak to friends and assemble trusted group for support team.

Signed
Timothy Jeffries

The Seed

Contradictory to much we've learned in academic and social pursuits, the sower and his care are more important than the seed and the soil. Day and night, the good farmer cultivates his field to reap a good harvest, but the weight of his care assumes the seed can only produce after its kind. We have wrongly assumed the seed in the ground (the gene in the human body) is the determining factor as to what the fruit (the human being) can become. What the scientific community has known, and is now making known, is this: our genetic structure (our seed) is a set of movable blocks akin to those in a child's playground. It is then the hands of the user, the creativity, and the nutrients given that ultimately forms what at first presents itself as preformed.

This is the taking of your genetic predisposition and repositioning it to engage a path none of your progenitors have ever engaged. You are redirecting a seed that took generations to get directly to you. You are disarming the assumption that genetic birth resists genetic deviation. You realize that while our genetic structure is written in pencil, our epigenetic lifestyle rewrites it in pen.

Geneticist Richard Lewontin in his book, *The Doctrine of DNA*, pulls the covers back when he writes, "While it is often said that DNA produces proteins, in fact, proteins (enzymes) produce DNA. When we refer to genes as self-replicating, we endow them with a mysterious autonomous power that seems to place them above the more ordinary materials of the body. Yet, if anything in the world can be said to be self-replicating, it is not the gene, but the entire organism (body) as a complex system." Humans reproduce in whole, not in part. Lewontin wisely views genes as footprints, which by default places our focus on the larger foot making the print.

This harkens a structural point. The body and its functions and flexibility stand counter-factually against other firm observations in earthly structures. The seed is less influential in creating the fruit. The real influence is what the seed is exposed to, is fed, is supported by, and how it's cultivated. Maybe this is how many from impoverished gutters of society can rise to heights not attained by those born in elevated places. Maybe this social phenomenon has its root in genetic truth. The teacher, who chooses to see more than a crack baby, can, through her encouragement, grow a senator. Her foot can overcome the footprint. This argues for the malleability of genes. Our genetic predisposition is our leaning, but our friends, family, and food is our nailing.

This book burrows a particular thought. The greater your understanding of cellular life, the greater control you'll have over your own. We have leveraged simple stories to highlight complex realities. While the polymorphic nature of teaching allows one set of principles to be communicated in millions of ways, the most effective form the world over is simplicity. The geneticist and the high school dropout must bow at its feet. In this book, you have the root; you have been given the intellectual gene to create the life you want, and now, as with all genes, what you do with it is up to you. It is that simple.

The Wall that Held

You needn't look at the florescent orange vest to know his line of work. Construction was all over his hands. He was just sitting there on the edge of a marble bench meant for viewing not resting. Unshaven and seemingly in his mid-fifties, there was something about him that wouldn't let me look away, not even for a second. I drew closer as the line I was in began to move. He seemed oblivious to the hundreds of people around him. The random conversations, children crying and security nudging us to keep moving didn't register with this man. He was in a trance. I turned the last corner and was placed directly in front of the person I'd been a fan of for the last ten minutes. He looked up, directly at me. He squeezed a smile which broke the tears dammed in his eyes. I caught my breath and looked around to see if anyone could see what I was seeing. This mountain of a man was crying in the most honest way I can express. It was then I knew this place was going to change my life forever. I couldn't really move until a loud but respectful instruction was tossed my way.

"Keep moving please."

I snapped out of it and fell back in line, frequently turning back around to see this intriguing construction worker crying on a slab of marble. As we got closer to our destination, it got quieter and quieter. Hundreds of people were standing around, but none of them made any unnecessary noises. Men and women in wheelchairs reached over the edge, rubbing their fingers in the grooves of the names forever etched in the facade. Seeing the 9/11 memorial on television was impactful, but being here was life-altering. There is a waterfall that stretches down into a seemingly black hole representing the serenity, respect, and honor for those buried underneath the towers, whose bodies were swallowed by the earth. Even the security personnel who are there day in and day out have a look on their faces of disbelief these many years later. Circling around to the enclosed memorial building, an attendant informed me that going inside would permit me to not only see artifacts from that infamous morning, but I'd be able to hear recordings from people calling their family members from the towers while the towers still stood, but it was the very last thing the attendant said that intrigued me. She said,

"You can even go underneath the towers."

I purchased my ticket and rushed into the building. This was a museum-like sanctuary built like the Smithsonian. As I wound my way to the bottom floor, I came upon a large tower of metal. This was the last piece of the towers seen in so many news reports over the years. The brave firemen kept this piece standing, using it as a wall to post notes to friends and family who never left the spot. I was walking toward the metal tower where so many had gather when I heard a whisper from my left. One of the tour guides was standing all alone saying something. I moved closer.

"The wall held."

"I'm sorry sir?"

"The wall held," the guide said again.

Right then, he pointed behind him at a gigantic structure, the type of structure you can walk right by and never see unless you look up. I was frozen in place. The wall looked tattered and torn. There was mold and mildew. Steel covered pipes protruded out of the wall in an orderly fashion every twenty feet or so in straight lines slightly pointing upward.

"Do you know the story?" The guide asked me.

"I do not. What is this?" I asked.

"Remember this name," the guide said.

"Arturo Lamberto Ressi di Cervia. He was the engineer who supervised the building of this wall you're standing next to over forty years ago. He was also the man who saved tens of thousands from dying on September 11th." The guide could tell I knew nothing about what he was referring to and was courteous enough to grant me a summary of the greatness towering over me. He continued.

"In the late 1960s the Hudson River, a 315-mile natural phenomena that starts at the Adirondack mountains and runs through Manhattan, had to be dammed so the World Trade buildings could be constructed. Mr. di Cervia, for some reason, refused to be satisfied with meeting the engineering requirements. He kept saying, 'It has to be stronger.' His team would pour twenty feet of concrete in width and seventy feet deep, but he kept insisting, 'It has to be stronger.' In the end, the wall known now as a 'slurry wall' was built several times stronger than engineering and science required. No one could imagine how incredibly insightful Mr. di Cervia was," the guide said.

By this time, I was beginning to understand what he was trying to say, the guide continued.

"On September 11th, millions of tons of pressure from the falling buildings should have breached Mr. di Cervia's wall. If the wall would have cracked you would have had billions of tons of waters flooding the subways of lower Manhattan. These subways have tens of thousands traveling underground at any moment of the day. These travelers would have been killed instantly. The toll could have risen to twenty or thirty thousand on that day, but the wall held...the wall held."

So much can be said about this. The wisdom of doing more

than necessary is astounding. The positive results are sobering. Mr. di Cervia built a wall of hope right in the middle of Manhattan having no idea he would be the last line of defense more than forty years later.

One day, things will fall around you. One day, the only thing that will permit your survival in times of "war" will be how you prepared in times of peace. Within our context and purpose of this writing, the enemies that may threaten your wall will be sickness and disease, cognitive decline, and physical decay. I have one last question, how strong is your wall? Are you building today a wall of health strong enough to handle surprise attacks? Are you stacking the bricks of eating, sleeping, breathing, motivating, supporting, resilience, motion, and emotion to handle the pressure that will one day attempt to break you?

Standing in front of that wall was sobering and yet it presented an opportunity to take its lesson and start today to prepare for tomorrow. I wish for you health, and health unmatched. Build your wall—build it today. May all your storms be weathered, and all that's good get better.

Ann-Louise Johnson, IFMCP, RN

DEDICATION

My clients. Thank you for holding on until I found you.

My parents. Dad, thank you for blessing me with your mantel of courage and adventure. Mom, thank you for the incredible gift of reading, and the hope reading grows.

My siblings. Perry and Priscilla, my brother and sister, and to the nine hearts that didn't make it to this life, but made it into my heart. Forever isn't long enough to love you.

My readers. With this book in hand, light the eternal flame within you. The flame of hope, yes, The DNA of Hope.

ACKNOWLEDGMENTS

With marvel and joy, I acknowledge the following professionals: Rita M. Rhoads, CNP, MPH; Anna B Morris, Health Coach; Deborah A. Everett, longtime friend and mentor; Arlene Weaver, hero and friend; Jeffery Bland, PhD; Peter D'Amaro, ND; David Jones, MD; Ben Lynch, ND; Bernarda Zenker, MD; Mark Houston, MD; Dale Bredesen, MD; Robert K. Naviaux, MD, PhD; and Andreas Wagner, PhD. To my writing coach, Dennis Ross, III for making my thoughts special, to graphic designers Anna Redos, Alexa Rearick, and Matthew Clark, thank you.

INDEX

A ability, Fogg Behavioral Model, 53–54
adenosine diphosphate (ADP), 74–75
adenosine triphosphate (ATP), 74–76
Adventurous Health, 144–148
alcohol addiction, 86–87
Alzheimer's disease, 195–196, 198–208
apolipoprotein E (ApoE) gene, 176–177
artificial neural networks, 6–7
aspartame, 82–84
asthma, breathing and, 161–162
autonomy, of human body, 80–81

B Baker, Sidney MacDonald, 15–16
Barabási, Albert-László, 126
belief
    hope and, 16–18
    resilience and, 184
    support and, 133–136
Bland, Jeffrey (Dr.), 192–193
breath and breathing
    cognitive function and, 207–208
    as genomic trigger, 11–12, 157–172
    hope and, 164–165
    intentionality of, 166–167
    mechanisms of, 159–160
Bredesen, Dale, 203–205
Buck Institute for Research on Aging, 203–205
butylated hydroxytoluene (BHT), 83
butylated hydroyanisol (BHA), 83

C cancer patients, support for, 144–155
Candle Problem, 25–26
cells and cell network
    cell phone model of, 10–11
    complexity of, 6–7
    creative influence over, 4–5
    seed model of, 98–105
    structure of, 28–31
    Winston narrative of, 105–116
cervical incompetence, 59–61
challenges, support and, 139

chromosomes, harmony in, 27
cognitive bias, functional fixedness and, 25–26
cognitive disease, management of, 200–208
collective, support and power of, 133–136
connections, in breathing, 168–169
core motivators, 54
CSLO gene, 28–31

D delusion, disease and, 191–193
dementia, 195–196, 198–208
di Cervia, Arturo Lamberto Ressi, 228–230
dimming, of suffering, 173–174
disease, definition of, 189–194
The Disease Delusion (Bland), 192–193
disproportional elongation, 78
dissonance, harmony vs., 26–27
diving, breathing and, 166–168
DNA, structure of, 28–31
DNA of Hope process, 209–223
doctor-patient partnership, 44–50
The Doctrine of DNA (Lewontin), 224–226
Duncker, Karl, 25–26

E electrons, in food, 69–73
emotion
    food and, 131–132
    as genomic trigger, 11–12, 119–132
    hope and, 119–128
    motion and, 129–131
emotional intelligence, 130
energy
    food as, 74–75
    sources of, 69–70
environment
    genetics and, 126–127
    health and, 45
epigenomic change, nutrition and, 85–89
excitatory amino acids, 82–84
excitotoxicity, 82–84
exercise, 129–130, 179–180, 206–207
exons, 28–29

F fiber networks, data transmission over, 5-6
Field to Plate movement, 79
Fogg, BJ, 53-54
Fogg Behavioral Model (FBM), 53-54
food
    cognitive function and, 201, 206-207
    emotion and, 131-132
    as force, 78-81, 91-93
    as genomic trigger, 11-12, 65-93, 212-222
    healing properties of, 49, 67-81, 148-149
    as hope, 90-93
    language around, 74
    resilience and, 180
    sleep and, 180-181
Food and Drug Administration (FDA), 84
force
    food as, 78-81, 91-93
    hope as, 119-128
friendship, 126-128
FTO genes, 126-127
functional fixedness, 25-26
functional medicine, 49-52
    cognitive decline and, 205-208

G genetics, environment and, 126-127
genomic kitchens, 79
genomic triggers, 26. *See also*
specific triggers, e.g., food
    breath, 11-12, 157-172
    emotion, 11-12, 119-132
    food, 11-12, 65-93
    as health collective, 158-159
    hope and, 190-194
    identification of, 11-14
    motion, 11-12, 119-132
    motivation, 11-12, 21-34
    resilience, 11-12, 173-194
    sleep, 11-12, 95-98
    support, 11-12, 133-155
genotype, 126
glutamate, 82-84
The Grey (film), 42, 44
ground thinking, 90-93

H habitual greatness, 53-55
harmony, motivation and, 26-27
health
    images of, 12-14
    sleep and, 95-97
    support and, 175-178
health profit, hope and, 190-194
Heal with a Meal program, 89, 214-215
hope
    food and, 90-93
    as habit, 51-52
    medicinal properties of, 9-12
    properties of, 119-128
    support and, 139-155
    transferability of, 136-139
Hope Creed, 210
Hope Network, 54
Hope Philharmonic, 19-24, 31-32

I "The Impossible Dream," 117
individuality, redundancy and, 15-16
integrity, resilience and, 185-187
introns, 28-29

J Jeffries, Timothy, 209-223
Jobs, Steve, V-VI
John, Tommy, I-IV
Johnson, Perry, 66-68

K Khan, Hazrat Imyat, 171-172
King, Martin Luther Jr., 115-116
Kummerow, Fred, 84

L language, hope and, 121-122
Lewontin, Richard, 224-225
lifestyle, health and, 13-14
living regimen, pathways to, 9-10
logic, belief and, 16-18
loss, acceptance of, 60-66
love, health and, 175-176

M Mate, Gabor (Dr.), 191-192
medicine
    nature *vs.*, 8-10
    nutrition and, 71-72
memory, 51, 195-196
Metabolic Enhancement for
NeuroDegeneration (MEND), 203-205
microwaved food, 75-76, 215-216
mitochondria doubling, 121, 166-170, 207
Molecules of Emotion (Pert), 131-132
motion
    as genomic trigger, 11-12, 119-132
    health and, 129-130
    hope and, 119-128
    resilience and, 179-180
    stillness as, 131-132
motivation
    breath and, 160
    Fogg Behavioral Model, 53-54
    as genomic trigger, 11-12, 21-34
    for health, 42
    hope and, 24-28, 35-36, 139
    in nature, 27-31

resilience and, 183-184
multi-core fiber networks, 5-6
mutable purpose, of food, 78

N nature
diversity in, 197-198
motivation in, 27-31
self-healing and self-correction in, 67
traditional medicine vs., 8-10
Neeson, Liam, 42, 44
negative resistance, 188-189
network analysis, 126-128
neurons, 82-84
neurotoxins, 82-84
The New England Journal of Medicine, 126-128
9/11 Memorial, 227-230
nutrition, 71-72
  disease and, 86-89

O Okene, Harrison, 36-41, 51-52
Okinawa Institute of Technology, 6-7
On the Trail (Johnson), 193-194
Operation Hope, 211

P pain, perception of, 42-44
patient-doctor partnership, 44-50
perspective, hope and, 158-161
Persuasive Tech laboratory, 53
Pert, Candace (Dr.), 131-132
phenotype, 126
population of the air, 89-93
Porter, James, 100-116
public outcry, 201-202

R regularity, of food, 78
requirement, food as, 78
resilience, as genomic trigger, 11-12, 173-194
resistance, resilience and, 187-189
retirement, sickness and, 44-45
Riken Research Institute, 7
RNA, 28

S seed model of cells, 98-105, 224-226
Siamese approach to illness, 130-131, 220-222
sleep
  cognitive function and, 207-208
  as genomic trigger, 11-12, 95-98, 212-222
  resilience and, 180-181
social networks, 126-128

statistics, limitations of, 77
stillness, as motion, 131-132
storm cells, 158
support
  cognitive function and, 205-207
  dementia and Alzheimers and, 200-208
  as genomic trigger, 11-12, 133-155, 212-222
  health and, 175-178
survival, hope and, 1-2
synthetic trans fats, 84-89

T telomeres, 191-192
tertiary butylhydroquinone (TBHQ), 83-84
thinking
  hope and, 17-18
  mind's appetite for, 195-198
touch
  healing properties of, 22-24
  transmission of hope through, 11-12
triggers, Fogg Behavioral Model, 53-54
Trimble, Ben, 176-177

U Umansky, Maxim, 41

V veterinary science, 71-72
Vitamin K2, 170-171

W water, chemical bonds in, 27-28
wellness
  hope and, 14
  sleep and, 97-98
white sugar, 84

"Hope is an intangible force. Through story and brilliant examples, Ann guides us through how significant hope truly is for you and your genes."

- Ben Lynch ND
Author of *Dirty Genes:*
A Breakthrough Program to Treat the Root Cause of Illness and Optimize Your Health

"The DNA of Hope is a profound commentary on how to regenerate our health, something useful for everyone today. Beginning at the cellular level, Ms. Johnson speaks of using our DNA (genetics) as a stepping stone to understanding how to live fully as we eat and change our lifestyle to express our phenotype. This is a must-read for anyone experiencing health problems or has lost hope. Her views on post-operative healing were so eye-opening. How diet and exercise can hasten our recovery was amazing."

- Rita Rhoads MPH, CRNP, CNM
Founder
Integrative Health Consults, LLC

ANN-LOUISE JOHNSON

Scientist, adventurer, medical professional, and internationally known expert, Ann-Louise Johnson is on a mission to teach people how to age fiercely. A registered nurse of more than forty years, and a certified functional medicine practitioner, Ann leverages intense personal narratives to inspire her clients to live longer, stronger, and with greater adventure than they could ever imagine.